Math and Magic in Camelot

Lilac Mohr

This book is dedicated to my parents
who taught me that:

there is profound
beauty and magic to
be discovered in the world
of math and science,

poetry provides
the perfect soundtrack
to life,

and smart children
are never bored!

Letter from the Author

Dear Reader,

If you've read *Math and Magic in Wonderland*, you've already met Lulu and Elizabeth and joined in as "*Mrs. Magpie's Manual*" led the twins on a grand problem-solving adventure. If you're not familiar with the first book in this series, don't despair! The experience is akin to making new friends and, little by little, gathering the vignettes that form the tapestry of their lives.

If you meet Lulu, she is sure to show you the Vorpal Blade given to her by Mrs. Magpie. She will tell you about her upcoming coronation and how she will soon be the queen of Tulgey Wood (a province of Wonderland). Elizabeth, Lulu's twin sister, will inform you that *she* has no need for a crown, only knowledge.

"Intelligence and a love for all that glitters are not mutually exclusive," Lulu will retort. She'll point at the girls' overflowing bookshelves. Elizabeth may respond with a quote from her favorite poet, Emily Dickinson: "There is no Frigate like a Book." And now you've had a proper introduction to the heroines of this novel.

As you read the story, be sure to try the activities in each chapter's "Play Along" section (found in the back of the book). Additional resources for each chapter are also available at *http://learnersinbloom.blogspot.com*. Enjoy!

Sincerely,
Lilac Mohr

Contents

1. The Message

*"If you can fill the unforgiving minute
With 60 seconds' worth of distance run,
Then yours is <u>Wonderland</u> and everything that's in it
And which is more, you'll <u>earn your sword and crown.</u>"*[1]

The Pigeon tapped his foot impatiently as Mrs. Magpie read the poem aloud. The contents of The Message were not his concern. The Pigeon was not fond of poetry and had never heard of Rudyard Kipling. Even if he had been a bibliophile like Mrs. Magpie, the Pigeon wouldn't have questioned her motives for changing the ending of a perfectly good poem. Indifference was part of the job. Whether Mrs. Magpie was of sound mind to place the fate of Wonderland in human hands was none of the Pigeon's business either. And for this, he was glad.

The Pigeon tapped his other foot as Mrs. Magpie scrawled "49° N, 77.47° W " across a crisp white envelope and placed The Message inside. He watched with disdain as she added two metal charms to the parcel (*oh, the weight!*), sealed it with a drop of wax (*more weight!*),

---

[1] Rudyard Kipling, "If—", 1895.   Underlined changes by Mrs. Magpie.

and attached it to his leg with thick twine (*oooh, the indignity of twine!*).

"You're about to meet two exceptional young ladies," said Mrs. Magpie as she tightened the knot. The Pigeon was silent, apathetic. If his genetics had allowed it, he might have yawned as a sign of disinterest. Mrs. Magpie studied the Pigeon's demeanor and nodded. *The Message was too important.* "Stay warm," she called as the Pigeon took to the sky.

The journey was long but uneventful. With biology and instinct taking care of the navigational details, the Pigeon had the luxury of operating on auto-pilot. He had plenty of time to think. Deciding that appeasing his employer was never a bad idea, the Pigeon mentally rehearsed the grand delivery of The Message. After finding the "exceptional young ladies", he would swoop gracefully over the water and land with a ceremonious bow at their feet. *Mrs. Magpie will be pleased to hear these details, no doubt. Being awarded a medal for heroism would not be out of the question.* The Pigeon would have smiled to himself if his biology had allowed it (*blasted biology!*).

It was morning when the messenger arrived at Lake Despinassy. An unusual tingling in the Pigeon's magnetite-charged beak was the first sign that something was amiss. It was immediately followed by another new sensation – disorientation. The needle of the Pigeon's internal compass spun wildly, and he momentarily fumbled in mid-air. Like any good ship-captain, the Pigeon quickly switched to manual steering and righted himself again. *What a strange place this is*, he thought. The Pigeon looped around the lake, looking, hesitating.

Two human girls stood like ghosts at the margins of the lake. They were staring bitterly at their reflections in the clear water. *Surely these frail shadows, pale and shivering, could not be the intended recipients of The Message!* On the chance that the girls, too, were affected by the anomaly in the Earth's magnetic field, the Pigeon decided to investigate further. He alighted in a nearby tree to listen.

One girl resembled a Mourning Dove – in mood or in physique the Pigeon wasn't sure (*perhaps both*). "Tomorrow's the day," she cooed solemnly. "Goodbye hopes! Farewell dreams and aspirations!" The girl placed the back of one hand upon her brow and the other upon her heart. She fell to her knees.

The messenger Pigeon was not one for empathy. If he had not been hindered by certain limitations of his class, the Pigeon would have rolled his eyes. Instead, his gaze shifted to the other girl.

This one resembled a Rock Pigeon like himself, *which undoubtedly meant she was the more sensible of the two.* This small human, "Cher Ami" he named her, knelt beside Mourning Dove and rested a thin hand on her shoulder. She picked up a rock and tossed it into the water. The ripples tore at the reflections of the trees with the precision of dominoes- systematically, without prejudice.

The Pigeon leaned forward as Cher Ami spoke:

"I felt a Funeral, in my Brain,
And Mourners to and fro
Kept treading - treading - till it seemed
That Sense was breaking through -"[2]

The Pigeon's pupils dilated. He raised his feathers in agitation and decided that he'd heard enough. *Sitting here while the recipients of the Message were waiting elsewhere was a waste of time.* Just as the Pigeon motioned to take off, Mourning Dove began laughing hysterically.

"Look at how pathetic we are!" she yelled at the trees. "Sulking and whining. What would Mrs. Magpie think of us now?"

At the mention of Mrs. Magpie, the Pigeon froze, still crouched, wings extended. Had he been endowed with different genetics, he would have perked his ears.

"You're right," agreed Cher Ami, standing up. "Just because Dad decided to take the family on a spontaneous fishing trip to Canada-"

She had spoken the word "Canada" with contempt, which restored the Pigeon's initial assessment of her character. "Too cold here," he mumbled.

"-right when we were scheduled to return to Tulgey Wood for my coronation-" Mourning Dove continued her companion's sentence.

The Pigeon observed that this had not ruffled Cher Ami's feathers. In fact, she resumed the sentence as if the two girls were of one mind – one bird continuing another's song. "- it doesn't mean that Mrs. Magpie can't use her magic to find us."

"In fact," yelled out Mourning Dove, "that pigeon over there could be one of her messengers for all we know," She pointed directly at the Pigeon who, startled, lost his balance and fell out of the tree. The Pigeon

[2] Emily Dickinson, "*I felt a Funeral, in my Brain*", 1896

wobbled along the ground toward the two girls and lifted a leg so they could untie the parcel. He tried not to think of the indignity of this position. The Pigeon managed to bow his head (more in embarrassment than reverence) before flying off. The girls watched the bird until he was out of sight.

"*Hope is the thing with feathers*,"[3] whispered Elizabeth emphatically. She was the one that the Pigeon had named "Cher Ami".

Her twin sister Lulu nodded in agreement. The Pigeon could not have known what a misnomer the moniker of 'Mourning Dove' had been. If he had stayed long enough to observe Lulu's 'freeform dance of ecstasy', perhaps he would have understood. But the Pigeon, mission accomplished, was already on his 2,700-km journey home. There would be plenty of time to reflect upon (and to embellish) the morning's events. Back in Wonderland, the Pigeon would share his account of meeting not 'exceptional young ladies' but 'mercurial squabs'.

Given that the Pigeon was not fond of poetry, he had only a limited notion of the power of words. This was something he could not rightfully blame on his avian nature (as Mrs. Magpie, a member of the same class, was exceptionally well-read herself). The Pigeon was thus unaware, of course, that sometimes an offhanded remark (or even one that had taken thousands of kilometers to devise) could pierce the fragile bubble of complacency. It

---

[3] Emily Dickinson, "*Hope is the thing with feathers*", 1891

was thus that the 'mercurial squabs' controversy (as it was later named) flamed the fires of rebellion in Wonderland.

But for us, Dear Readers, this is where the Pigeon's story ends, and the story of Lulu and Elizabeth continues. Alas, patience is a virtue which human children and fledglings, alike, possess in limited quantities. For the twins, a morning of frenzied speculation melted into an afternoon of silent anticipation, which in turn fell slowly (*oh so slowly*) into an evening of torment. Now we find the girls sitting on their beds in the dark cabin negotiating the blurry line between enthusiasm and restlessness as they wait for morning.

Play Along: In this chapter's "Play Along" section, you'll learn more about homing pigeons and the heroic Cher Ami. You'll also make a compass as you explore magnetoreception. Join the fun on page 191.

2. Waiting for Wonderland

"Wonderland was only a dream," Elizabeth whispered as she turned her face to the window. Her tone was not confrontational, but pensive.

To Lulu, the words were an assault nonetheless. She sprang to her feet at once and reached for the small sword around her neck. Her other hand pulled out one of the metal charms that Mrs. Magpie had enclosed in the Pigeon's parcel.

Elizabeth smiled when she caught a glimpse of her sister's staunch pout reflected in the window. "No, not our Wonderland," she corrected. "I was talking about the one in Lewis Carroll's book. Maybe falling asleep will take us back."

Lulu's face relaxed for only a moment. She released the dagger and returned the charm to her pocket. Next, Lulu retrieved the worn envelope from the other pocket and grimaced. The Message itself was beautiful, wonderful, and more than welcome – 'a promise of springtime' was the best way Lulu could describe it.

Unfortunately, it was only autumn, and a long winter still loomed ahead. The agony of the wait had, hours ago, transformed the promised sword and crown into a sore and frown.

Lulu sighed loudly. "You can go to sleep if you'd like, but I intend to stay awake until spring arrives."

"Spring?" Elizabeth asked. She turned to face her sister. "Do you mean morning? That leaves you with-" Elizabeth consulted her watch. "-about nine more hours of moping."

Lulu smirked. "Who said anything about moping?" she replied. "Doesn't Mom always remind us that smart children are never bored?"

Lulu opened the pink notebook that was never far from her side and retrieved her favorite silver glitter pen from under her pillow. She began writing. "Mom is not the only family member who can write novels, you know. I'm going to be an author, too. Chapter 1: Waiting for Wonderland."

"You can't start your book with characters sitting around and waiting. The readers will be yawning in no time," said Elizabeth as a yawn escaped from her own mouth.

Lulu always took criticism in stride (usually by ignoring it). "The cure for ennui is a cup of tea," she read from her notes. "Sounds a bit like Confucius[4], don't you think?"

"It sounds exactly… like a tea commercial!" snickered Elizabeth. "Besides, everyone knows that the cure for boredom… is a poem!"

Lulu was pleasantly surprised to find that the delirium of sleeplessness had transformed her sister into a

---

[4] Chinese educator and philosopher (~500BC)

more playful version of her usually grave and skeptical self.

"Lulu Lovelace, I challenge you to a duel!" Elizabeth suddenly yelled out.

Lulu blinked in disbelief. *Had madness set in so quickly?* She placed a finger over her lips. If their parents or younger brothers awoke, the promise of Wonderland would vanish. "Shhh!" The broad smile on Lulu's face was a clear window to her delight. She grabbed for the Vorpal Blade around her neck.

Elizabeth watched her sister curiously. "Not a duel with swords-" she laughed, "- a battle with words. A Poetry Duel! The theme is sleep!"

Lulu let go of the Vorpal Blade, which fell back against her chest. She swung her glitter pen in the air. "That is the nerdiest thing I've ever heard... I love it!" Lulu scribbled frantically in her notebook.

"I hope you don't include this in your novel," said Elizabeth smugly. "If you've put half your readers to sleep with all the sitting around and talking, a poetry duel is sure to do away with the other half."

Not madness, after all. Sweet, cynical Elizabeth is back, thought Lulu as she continued writing.

"On the contrary, Dear Sister," she finally responded (as this is how the Lovelace sisters spoke to each other - the product of reading classic novels). "Maybe we'll start a new trend; Young people all over the world will gather secretly in the veil of night to do battle... with Victorian poetry! I can see it now..."

Elizabeth shrugged. "I'll go first - with a poem by Emily Dickinson."

The Elizabeth I know and love is back without a doubt, thought Lulu. She must have said the words aloud because Elizabeth was glaring at her.

"I know you think Emily was just a kooky lady who dressed in white and loved nature more than people, but you will have to admit that this poem is the winner. Remember that the theme is *sleep*."

Lulu, who found Elizabeth's description of Emily Dickinson sufficient, pursed her lips and listened.

"Will there really be a "Morning"?
Is there such a thing as "Day"?
Could I see it from the mountains
If I were as tall as they?

Has it feet like Water lilies?
Has it feathers like a Bird?
Is it brought from famous countries
Of which I have never heard?

Oh some Scholar! Oh some Sailor!
Oh some Wise Man from the skies!
Please to tell a little Pilgrim
Where the place called "Morning" lies! "[5]

Elizabeth let the words of the poem linger in the air. The silence was her version of a dramatic ending – an appropriate homage to Ms. Dickinson's genius.

"Now *that's* a poem you don't understand until you've battled with insomnia," said Lulu, breaking the silence.

"There's only one thing I dislike about it," Elizabeth admitted. "Emily gives equal value to the advice of both the Scholar and the Sailor -"

"And don't forget the Wise Man from the skies," Lulu interjected.

---

[5] Emily Dickinson, *"Will there really be a 'Morning'?"*, 1896

"A scholar is educated and, well, a sailor..."

"Call me Ishmael,[6]" said Lulu, reciting the opening lines of *Moby Dick* in her surliest sailor voice.

"Exactly!"

"Think not, is my eleventh commandment; and sleep when you can, is my twelfth.[7]" Lulu continued, now quoting the book's jovial second mate.

"Point taken," said Elizabeth surveying the paintings of maritime images on the cabin walls. "What could a sailor and a scholar possibly have in common?"

Lulu took off her glasses and rubbed her tired eyes. "It's like asking how a raven is like a writing-desk[8]. There's no answer- " It was at that moment that she spotted a small green glow in the corner of the room.

N. Wake, the leader of Mouse Troupe 4.6692, knew exactly how a Raven was like a writing desk (and did not hold a favorable view of either). 'Antediluvian' was one word which came to mind. She also knew how a certain Mouse was like a jellyfish. The word was 'brilliant', of course. N. Wake laughed at her witty pun - a joke for which she would never receive credit as that information was strictly classified as D.N.A ("Do Not Announce").

Leader, spy, messenger, and master of bio-florescence, N. Wake was ready for her first official mission. She hoped to prove as smart and brave as her namesake, Nancy Wake, the decorated World War II spy and the original 'White Mouse'.

---

[6] Herman Melville, "*Moby-Dick; or, The Whale*", 1851

[7] Herman Melville, "*Moby-Dick; or, The Whale*", 1851

[8] A riddle posed by the Mad Hatter in Lewis Carroll's book "*Alice in Wonderland*", 1865

Sometimes courage comes riding into existence upon the back of a powerful tidal wave. More frequently, however, it takes the form of tiny vibrations. When the tingling began- first in the Mouse's feet and soon undulating up through her body to the tip of her green head- it was time. N. Wake motioned to the other three members of her team... and darted.

Lulu, her glasses off, saw only a flare of green light shoot across the room. Elizabeth saw the Mice clearly (which may or may not have been a good thing). She jumped up on her bed and held a pillow over her mouth.

"What was that?" asked Lulu excitedly.

As the urge to scream subsided, Elizabeth removed the pillow. Her heart continued to race, and the battle between scientific curiosity and fear still raged in her head. "Four mice. Three brown. One green. They dove under there," she said, pointing to a shaggy blue rug. The best way to resolve an internal battle, of course, is to state the facts. Facts never pass judgment and never take sides (as they leave such unpleasantness to others).

Lulu, her glasses firmly in place, leaped upon the rug and pinned it down with both hands and both feet. "Search for lumps," she called out to her sister. One look at Elizabeth's terror-filled eyes and Lulu began a visual inspection of the rug for herself. Although there were no discernible lumps in the dense shag, four small mice could easily vanish beneath the thick material.

Lulu gingerly lifted a corner and slowly peeled the rug off the floor. Instead of mice, she found a wooden panel, crudely constructed from planks of wood that were a different shade from the rest of the floor. Each one had a word burned into it. "Mustard, Anise, Garlic, Ice Plant, Chives," Lulu read aloud.

"A spice cellar?" asked Elizabeth who had temporarily overcome her fear of rodents to join her sister on the floor (as scientific curiosity trumps fear every time!). "Why would anyone need that?"

Lulu slid her fingers into the wood's knot holes (through which the mice had undoubtedly escaped). She quickly lifted the panel. A different panel lay underneath - an older one. Its wood was warped and faded, looking as if it had suffered extensive water damage. Lulu squinted as she tried to read the faded red letters that were painted on the wood. "M-A-G" Lulu began. "Magpie! What else could it be?" she exclaimed.

"Magnolia?" offered Elizabeth. "Are magnolia flowers edible?"

"It has to be Magpie - I'm sure of it," said Lulu, wiggling her shoulders back and forth in a little celebratory dance. "Mrs. Magpie wanted us to find this place. It's the way back to Wonderland!" Without hesitation, her fingers removed the second panel. Lulu stuck her face into the opening.

"It's too dark," she said, turning to her sister. "Hand me the flashlight."

Elizabeth could not find the right words with which to protest. *Trespassing, irresponsible,* and *dangerous* all came to mind as she grabbed the flashlight from the nightstand and held it toward Lulu. "*Like one in danger; cautious, I offered him a crumb[9],*" she whispered, remembering another Emily Dickinson poem. While the element of danger was evident, 'cautious' was not a word

---

[9] Emily Dickinson "*A Bird came down the walk*", 1891

Elizabeth would use to describe her sister. *Am I the bird, about to unroll my feathers and row home?* she wondered. Elizabeth's body suddenly felt uncharacteristically light, almost feathery. *Could it be a sense of adventure?* A loud thump snapped Elizabeth from her daydream. Lulu was gone.

Elizabeth peered into the trap door's opening. A small beam of light from the flashlight was bouncing around the walls. "Jump in!" Lulu yelled up at her. Elizabeth, completely out of character, unrolled her wings.

The landing was soft. "It's a bed!" explained Lulu, preemptively answering her sister's question. A thin beam of pale blue moonlight fell through the cellar entrance. As Elizabeth's eyes adjusted, the dark silhouette of an enormous wooden bedframe slowly took form like a ship arising from the fog.

The ship analogy seemed even more suitable to Elizabeth as her eyes chased the beam of Lulu's flashlight around the room. It bounced wildly across the walls, each time landing on a different picture. A large, vibrant acrylic painting was illuminated one second, and a small soft-toned watercolor the next. The light bounced across oil paintings, photographs, and pencil sketches.

Like the rest of the cabin, the images all depicted maritime scenes. Some of the pictures showcased large schooners with three or four tall masts and a whole army of sails. Others contained smaller sailboats, long flat cargo vessels, or tiny fishing boats. There was even an oil painting of a Viking ship. Lulu paused the flashlight's beam on this one longer than the others. Elizabeth just stared, grasping for connections that might make sense of the objects in this bizarre cellar.

Lulu began jumping up and down on the bed. "Look behind you," she beckoned with an excited smile. Elizabeth spun around and gasped. On each side of the headboard loomed the wooden head of a dragon. Both imposing dragon-heads were intricately carved and painted so that their eyes twinkled mysteriously and each wide mouth lay closed in a knowing smile. The letter 'N' was burned onto one dragon's neck and 'W' onto the other's.

The girls stared at the beasts in quiet awe for a very long time. Lulu was the first to arouse from the dragons' trance. She directed the flashlight's beam to the bed's headboard. Embedded at its center, was an image which both girls immediately recognized. Lulu reached into her pocket and clutched Mrs. Magpie's charm. "The trefoil knot," she whispered.

Elizabeth, who felt uneasy the moment she saw the monstrous bed and, to be honest, had been contemplating an escape plan just minutes ago, fell to her knees and pressed her fingers to the symbol. You see, Dear Readers, that although Lulu and Elizabeth often read of other times – older times - when honor was a prized procession, and a man's word was as good as gold, the girls were born into a world where promises were broken without consequence. There is no doubt that it was a beautiful world in which Lulu and Elizabeth lived – a world full of knowledge and opportunity and privilege. But it was a world, nonetheless, where words could be blown to the wind as easily as a weed casts off its seeds. Because of these circumstances, it was only here -in the mysterious cellar with the discovery of the trefoil knot symbol - that Mrs. Magpie's promise became tangible. It was only with that discovery that the promise of Wonderland turned real.

Elizabeth traced the trefoil knot with her fingertip, and a faint glow appeared around the symbol. *Was it a triboelectric effect?* she wondered. Elizabeth's finger continued moving around and around on its infinite path. As it crossed each of the knot's three intersections, the light surrounding the symbol intensified. Elizabeth turned her face away from the glare. A strange tingle crept up her arm, but she could not stop rubbing.

Pop! Creak. Grrrr. Elizabeth was thrown back onto the bed. Terrified, she scrambled to her feet and flung herself at the cellar's opening, jumping up with both arms outstretched. A ladder slid down silently in front of her. Its top was affixed to the trap door's opening, and its base rested on the bed. Elizabeth placed both hands on the ladder and was about to climb when her sister's voice gave her pause. "The dragons," Lulu was yelling. "Look at the dragons!"

Elizabeth looked. An eerie red light illuminated the two dragon heads so they appeared almost alive. Their mouths lay open, and a long tongue unrolled from each. Lulu was pointing at the headboard where a ruby was glowing at the trefoil symbol's center, but Elizabeth could not take her eyes off the dragons.

Forgetting the ladder, she approached the dragon with the 'N' on its neck, stared, cocked her head, and then moved to the one that bore the letter 'W'. Laying a shaky hand against the beast's great muzzle, she brought her face to the dragon's mouth. "It's a puzzle from Mrs. Magpie," said Elizabeth, her voice surprisingly steady.

Lulu moved next to her sister and peered into the wooden jaws. Inside the dragon's mouth was a dial surrounded by numbers. Lulu smiled. She shared her sister's enthusiasm for math and logic puzzles. The world of mathematics was comfortable and familiar, logical and reliable (as the twins had not yet encountered the abstract concepts of pure mathematics).

"The numbers around this dial go up to one hundred and eighty," noted Elizabeth. "And the other one goes only to ninety." With a quick glance into the other dragon's mouth, Lulu confirmed her sister's findings. "The Message-" continued Elizabeth without removing her gaze from the dragon's mouth. "The Message is a clue."

Lulu removed the envelope from her pocket and pulled out Mrs. Magpie's note. This time it felt different - as if she were seeing it for the first time. Lulu shined the light on the paper and stared once again at the words she had long ago committed to memory. "*If you can fill the unforgiving minute, with sixty seconds' worth of distance run-*" Lulu's eyes brightened. "Sixty seconds - "

"No, not the poem —the envelope," interrupted Elizabeth. "Look at the envelope!"

Lulu shined the flashlight at the front of the envelope. "49° N, 77.47° W," she read aloud. "This is too easy!" Lulu looked in the first dragon's mouth and stared at the dial with a puzzled expression. "It's already turned to 49," she said.

Elizabeth finally turned to look at her sister. "This one has already been set too," she said.

Lulu wrinkled her nose. "This is definitely too easy," she said, trying to hide her disappointment as she remembered why solving math problems for herself was so much more gratifying than just copying the answers from the back of the book. "I guess we're ready to fall asleep and wake up in Wonderland," Lulu said with an uneasy shrug.

Elizabeth only grunted in response.

Lulu was none too pleased either. "Give me a math problem, and I'll solve it. Challenge me with a riddle, and I'll figure it out. Send me into battle, and I'll rise to the occasion with my sword drawn and ready. But please, oh please, don't make me fall asleep," she pleaded to the dragons.

"Maybe Mrs. Magpie is teaching us a lesson in perseverance," suggested Elizabeth.

"Indeed," Lulu agreed with another sigh. She flopped dramatically onto the bed and began reciting the digits of pi ("more humane than counting sheep," she would always joke).

Elizabeth turned her back to the dragons and felt her body crumple onto the bed, almost involuntarily. The bed was not particularly comfortable. Its covering was a thick, scratchy burlap cloth, not the soft linen or warm wool she was expecting. *Part of the lesson*, thought Elizabeth as she closed her eyes. She listened to her sister's rhythmical counting and tried to relax.

Suddenly, Lulu stopped reciting and sat up. She pointed the flashlight at the foot-board. At first, she thought the wood was carved with a detailed design of some sort. Upon closer inspection, however, the girls saw that the carvings were names, initials, hearts, and short messages such as lovers may carve into a beech tree. "Who were these people?" asked Lulu as her fingers traced over the names. With her other hand, she reached for the dagger around her neck.

Elizabeth gave her sister an admonishing look. "I don't think Mrs. Magpie gave you the Vorpal Blade for vandalizing an antique bed."

"It's in the name of posterity," Lulu insisted. She reached to pull the dagger from its sheath and bumped the flashlight. The flashlight's beam landed on one of the footboard's messages. This one was carved inside a large heart. Lulu let go of the dagger and read the words aloud:

There are sleeping dreams and waking dreams;
What seems is not always as it seems.[10]

"Christina Rossetti?" both girls wondered at once, exchanging wide-eyed looks.

"It's from her poem *'A Ballad of Boding'*," added Lulu.

"Ms. Rossetti is the winner of our poetry duel," joked Elizabeth. She let out an involuntary yawn.

Lulu yawned as well, thinking of how fatigue turns one's external and internal self into a humongous contradiction.

"I had a better one," she said, blinking the excitement back to life in her eyes. Lulu pulled out her notebook. "It's *'The Bed Book'* by Sylvia Plath."

"You'll win in the irony category," joked Elizabeth, tapping the burlap bed-cover. "I supposed you want to share it."

"Nope. I can't."

Elizabeth, suddenly craving a poem as a baby awaits his lullaby, walked into her sister's trap. "You can't read it aloud? Why not?"

"Because it's still under copyright protection. I'll mention the poem's name in my book, but won't reveal its contents."

Elizabeth wrinkled her nose. "Your readers will be confused... or mad."

---

[10] Christina Georgina Rossetti, "*A Ballad of Boding*",1881

"Not at all. My readers are smart. And smart people are curious. And curious people are resourceful. And, besides, it's the right sort of poem - if you know what I mean."

"I'm afraid I don't," Elizabeth replied with a weary sigh (as she wasn't convinced that any work written after 1923 had much literary merit).

Lulu was waiting for this exact interaction. She took a breath and began reciting, not Sylvia Plath's poem, but one she herself had composed earlier in the day[11]:

> *"Most poems are pleasant*
> *And fine for reciting*
> *But the best poems are*
> *Much more exciting.*
>
> *Not just a sweet poem,*
> *Follow the beat poem,*
> *Tidy and neat poem,*
> *Stay in your seat poem,*
>
> *Instead...*
> *A poem with thrills*
> *A poem with chills*
> *A poem that turns you head over heels,"*

"Sounds dangerous," interrupted Elizabeth.
"Not necessarily..."

> *"The right sort of poem,*
> *(If you see what I mean)*
> *Is one that might send you*
> *Into a dream*

---

[11] Written in the style of "*The Bed Book*" by Sylvia Plath, 1976

Through a medieval forest,
Lush and green,
Where fairies transform you
Into a Queen!"

Lulu finished in a soft voice, a dreamy voice, the sort of voice that is soon met with heavy eyelids where reality and unreality coalesce into one... big... blur...

Play Along: Discover patterns in poetry, explore knot theory, learn about phosphorescent mice, and more. Join the fun on page 193.

3. Scholars and Sailors

Elizabeth Lovelace awoke from uneasy slumber to find herself facing a nightmare of a different sort. This one took the shape of a small green mouse perched on her chest. Elizabeth screamed and the Mouse (it was N. Wake, of course) dove beneath the covers.

"Are we in Wonderland?" asked Lulu drowsily.

Elizabeth heard her sister but did not respond until she was safely out of the cellar. "We're not in Wonderland," she called down. The rodent-encounter was fresh, and her voice was shaky. The world outside the window was still enveloped in darkness. "It's the same old vacation cabin in Quebec. Nothing's changed," Elizabeth said. The words filled her with an unexpected sense of comfort.

Lulu's head emerged from the cellar. "I thought I heard a scream."

Elizabeth cringed. "I'm not going down there again," she said with a shiver. "The bed is crawling with mice!"

"Mice? I'll investigate," said Lulu and once again disappeared into the cellar opening.

Dropping back on the enormous ship-bed, Lulu tucked the flashlight under her chin and drew back the thick burlap comforter with both hands. She would have been surprised to find a modern mattress on such an ancient-looking bed if she had not been more startled to discover that the center of the mattress was gutted. As Lulu dug through torn padding and exposed springs, her fingers wrapped around a small cardboard box.

Lulu brushed the loose fibers off the container and removed its lid. Five mice looked up at her. An abnormally large brown mouse sat in the center of the box. Surrounding it, were four smaller mice- three brown and one green. "A recessive gene," Lulu said aloud, her suspicions confirmed.

The brown mice remained frozen in their places. The green one swept its long tail back and forth across the paper that lined the container. When the green mouse moved aside, Lulu could see letters on the paper.

"I hope you don't mind," said Lulu to the mice as she gently removed the paper. The brown mice remained stiff as the paper was slid out from beneath them. The green one performed – *was it a pirouette?*

Lulu unfolded the paper to its full size and smoothed it out against her thigh. It was a handwritten note, surprisingly intact considering its most recent location. Lulu read:

Bed #270. Default: 49, 77.47

The coordinates to Magic
Should pose you no trouble.
Fear-not a life pelagic
As the distance will double.

"The coordinates to Magic?" Lulu asked aloud. The green mouse let out an audible sigh.

N. Wake looked over at Ms. Magnum, the spy-master to which she reported. Ms. Magnum, the larger brown Mouse (who bore no direct genetic relation to N. Wake as Lulu had assumed), stared at her with a look of vexation. N. Wake, in turn, committed her first thought-crime of her first official mission: questioning authority.

Ms. Magnum had communicated the mission directive clearly that morning: "Deliver the girls to Q without using your words." *Without using your words.* Here is where the thought-crime had first started to take form.

Words were N. Wake's greatest treasure. She could take no credit for the bio-florescence that identified her as a highly-trained genetically-engineered spy-Mouse (as it was the work of others). But words would always be her personal property. Words could accomplish acts beyond the capabilities of any spy gadget. To N. Wake, it was the infinite arsenal of words in her possession, not the unusual color of her skin, that was her defining feature. Words were everything, and without them, she was no longer a Mouse.

N. Wake had been excited about her first mission, but she also secretly stewed on the injustice of being deprived of her most powerful tool and greatest weapon. Perhaps this insolence was responsible for N. Wake's first big mistake (not to be confused with her thought-crime, as thoughts, even those filled with misconceptions, can never be properly classified as mistakes).

N. Wake had been careless. She had chosen the wrong girl – the one afraid of mice – and thus had inadvertently revealed the location of the safehouse. Later, there would be repercussions. N. Wake hoped that Q, the mastermind of the entire spy-Mouse operation, would be more forgiving than Ms. Magnum's cold stare implied.

Lulu brought her face closer to the box. Her eyes met melancholy Mouse eyes, and the communication was anything but mouse-like. "Thank you for all your help," whispered Lulu. N. Wake nodded. *Did a nod violate protocol?*

"Were you sent by Mrs. Magpie?" At this question, N. Wake shook her head vehemently, bitter that she was indeed 'sent' and was not operating on her own authority. *Were thought-crimes on par with treason?*

There was a long pause. Lulu never took her eyes off the Mouse. "What are the coordinates to Magic?" she asked softly. N. Wake fought the urge to speak. The words formed a hard lump in her throat. She could feel Ms. Magnum's piercing glare burning her cheek.

First, N. Wake ran a quick damage-control algorithm in her advanced brain. Next, she swallowed away the lump (along with her pride), let out a squeak, and scurried to the wall that pressed against the foot of the bed. Lulu followed the Mouse's path with her flashlight (although the bright green glow marked the way clearly).

The green mouse vanished into the corner of a wide painting that hung slightly askew on the wall. Lulu slowly scanned the picture with her flashlight. The upper left-hand corner of the canvas held a poem. Lulu recognized it immediately - "Sea Fever" by John Masefield:

"I must go down to the seas again, to the lonely sea and the sky,

And all I ask is a tall ship and a star to steer her by."[12]

Next to the letters, a full moon stood in command of a starry sky and a watery domain. Her moonbeams highlighted the white-tips of the frothy waves as they lapped at a handsome ship. Lulu's thoughts drifted to the romance of a vagabond life at sea (for she knew nothing of sailing nor the type of winds that produce frothy white-tipped waves).

As the flashlight's glow moved to the other side of the image, however, Lulu's heart began to thump with a sense of foreboding. In this region, a cavalry of dark clouds was charging the vessel. The ship was tilted almost horizontally, bracing itself against the waves, its torn sails flailing in the wind.

A piece of yellowing paper was pasted to this side of the canvas. It was written in the same handwriting as the clue Lulu had retrieved from the mouse box. Lulu did not recognize this poem.

> *"Woe to the sailor far out at sea,*
> *With only Polaris for company.*
> *Currents and winds whisper sinister plots,*
> *As you let out the rope and count up the knots.*
>
> *To know your position before a collision*
> *You need the precision of Harrison's vision.*
> *(This is a tip, not a derision)*
>
> *Crash!*
>
> *Your lips taste of salty brine*
> *As you gaze at the stars a final time.*
> *A sailor's lament as he hits the rock:*
> *Longitude depends on an accurate -"*

[12] John Masefield, "Sea Fever", 1913

The rest of the paper was torn. Quickly deciding against a vagabond life at sea (and realizing that poetry and art are often more romantic than reality), Lulu quickly replaced the burlap bed covering and called up to her sister.

"Elizabeth, come quickly! I found some clues!" At first, there was only silence upstairs, then footsteps, and finally Elizabeth's face appeared at the cellar entrance. Elizabeth had her arsenal of words loaded and ready, but she put them aside at the sight of the note in her sister's hands. Curiosity was squarely to blame, and Elizabeth knew it. She descended the ladder and thought no more of gnawing mammals nor other personages with sharp teeth.

Lulu handed the bed's instructions to Elizabeth and then shone the flashlight at the painting. "It's changed," she gasped. The clouds now completely obscured the moon, and the girls had to strain their eyes to spot the outline of the ship between dark sea and black sky.

Elizabeth had already skimmed the instructions. She pushed the paper into Lulu's hand and moved closer to the note that was affixed to the painting. "Harrison's vision!" Elizabeth exclaimed. Her expression soon grew grave, however, as she continued reading. Placing a shaky hand on each side of the painting, Elizabeth read the last line aloud:

> "*A sailor's lament as he hits the rock:*
> *Longitude depends on an accurate -*"

The wall groaned. Elizabeth let out a scream and sprung back as the drywall behind the painting cracked open. The large picture was the first to crash down onto the bed. More maritime paintings followed along with a

27

rain of plaster. Lulu and Elizabeth could only stare as the crack in the wall widened, and the two sides began sliding apart, sending down another deluge of drywall chunks. This time they fell with a clank, not on the bed, but on the floor. Before the girls could say a word, the wall heaved a heavy sigh and came to a stop.

The dust began to clear, and Lulu and Elizabeth found themselves facing a closet that spanned the entire length of what had been the wall. The girls brushed pieces of drywall from their clothing and stepped into the closet.

As Lulu scanned the space with her flashlight, the twins could see that the closet was full of clocks. Many of the clocks stood on shelves, their exposed gears spinning rhythmically. There was a grandfather clock on the floor with a giant swinging pendulum. Smaller clocks, resembling large watches, were hanging on the wall.

"Tokyo, New York, New Delhi," said Elizabeth, reading the neat labels next to the clocks. "South Cadbury Castle, Greenwich," she continued.

"Lake Despinassy, Magic," read Lulu.

The twins looked at each other in surprise. "Magic?" Elizabeth asked. "Is it a real place?"

Lulu shrugged. "It is not down in any map;" she quoted in her sailor's voice, "true places never are."[13]

Elizabeth's eyes twinkled. "I know why the clocks are here," she said. "I know what we have to do."

Lulu turned to face her sister.

Elizabeth took a breath and continued. "To find your longitude, you need a clock - and an accurate one at that."

---

[13] Herman Melville, *"Moby-Dick; or, The Whale"*, 1851

"The missing word in the poem!"

Elizabeth nodded. "Let me tell you a little story about a man named John Harrison."

"If you can wait and not be tired by waiting,"[14]
-Rudyard Kipling

Harrison looked up at the night sky. He turned his back to the moon and found Polaris winking at him. Harrison grimaced. He had endured the North Star's taunting glare for forty-seven years. For forty-seven years, the North Star, every sailor's best friend, was a constant reminder of how easy - how uncontested - it was to find a ship's latitude. But today - today it was Harrison's turn to gloat.

"Where were you on that fateful October night in 1707 when four of Queen Anne's Royal Naval ships crashed into the Isles of Scilly?" Harrison yelled at his celestial tormentor. The night air sent him into a coughing fit but did nothing to diminish his enthusiasm. "Where were you when 1,550 sailors were sent to their watery graves?" Harrison suppressed another cough and shook a fist at Polaris.

In 1714, when Parliament passed the Longitude Act, everyone was convinced that it would be a scholar who claimed the prize at hand. Only a scholar could devise a technique to determine the longitude of a ship with enough accuracy to take home the prize. Even the members of the Longitude Board were waiting for a brilliant astronomer or mathematician to calculate the

---

[14] Rudyard Kipling, "*If—*", 1895

horizontal line, or meridian, of a ship's current position using the moon.

Imagine their surprise when they were contacted by a common carpenter and clockmaker. Harrison smiled to himself. He pulled an envelope from his pocket and ran his wrinkled fingers gently across it. "Forty-seven years," he muttered. "I knew from the beginning that it was just a matter of time."

From the day Harrison began pursuing the Longitude Problem in the 1720's, his approach had differed from the scholars'. A time-based calculation for longitude, or the east-west position of a point on the globe, was already known. Every hour, the sun moves 15 degrees across the sky (360 degrees in a circle divided by 24 hours in a day). Find the difference between the time where the sun is highest in the sky in your current location to noon on the clock that you set before leaving port. Multiply the number by 15 degrees per hour, and you've calculated the number of degrees east or west that your ship has traveled.

Harrison thought about the longitude calculation. "Easy enough?" he asked aloud. "Therein lies the rub!" The dry chuckle that came from his throat reminded Harrison that he was no longer a young man.

Developing a sea-worthy chronometer that would not lose its accuracy while subjected to the movements and extreme conditions onboard a ship took Harrison a lifetime. Design, test, improve, repeat. "Repeat and repeat and repeat once again," Harrison recalled. Indeed, he would repeat this cycle over and over, each iteration bringing him closer to the end of the race - closer to the prize. Now the heavens bore witness to his victory. The heavens and the little green mouse hiding behind his leg.

Harrison tenderly stroked the envelope in his hand. King George III had encouraged Parliament to pass a new Act awarding Harrison the prize money. This letter was proof. "Forty-seven years," whisper Harrison. "Forty-seven years." A gust of wind echoed his sentiments. *Recognition at last!*

"That puts things in perspective," said Lulu when Elizabeth had concluded the story. "Waiting a couple of hours for Wonderland is not a test in perseverance."

The sisters studied the row of clocks. "Looks like it's 4:15 at the Royal Observatory in Greenwich and 1:35 in Magic, wherever that is," said Elizabeth, pointing first to the grandfather clock and then at a clock that resembled a large watch.

"I don't think these clocks are accurate," said Lulu, trying to hide the disappointment in her voice. "The Lake Despinassy clock says it's 11:05, but my watch says 12:15. It's more than an hour slow."

"Clocks do lose their accuracy over time," conceded Elizabeth. "Even Harrison's H4 clock - the most accurate clock in the world at the time - lost about 24 seconds a day compared to solar time. Those seconds can add up quickly."

"Did you say, 'compared to solar time'?" asked Lulu. "That's the answer! My watch doesn't tell us the solar time because time zones group entire regions together."

Elizabeth pulled a calculator from her pocket. She found the difference between the Lake Despinassy clock and Greenwich Mean Time (the clock labeled 'Greenwich'), converted the difference to hours, and multiplied by fifteen. "What were the numbers on Ms. Magpie's envelope?" she asked.

"49° N, 77.47° W" replied Lulu from memory.

Elizabeth looked at her calculator and smiled. "The clocks are accurate. We have a way to calculate longitude!"

Lulu let out a hoot of glee. Computing the longitude to Magic was just a matter of repeating the process using the clock labeled 'Magic'. "Forty West," said Elizabeth after completing the calculation[15]. "But that's only one of the coordinates to Magic. To get the latitude, we need to find Polaris, the North Star. But first, we'll build a sextant-"

"I think I know the solution," said Lulu excitedly. "Mrs. Magpie's message quoted a poem by Rudyard Kipling titled '*IF*'. Maybe Kipling has the key to this riddle, as well. Remember the location of the shipwrecked mariner in '*How the Whale Got His Throat*'?

"Fifty North and Forty West!" yelled out Elizabeth.

"And that's magic," added Lulu in a whisper (the way Mom always read it). "Kipling said so himself. Let's try it!"

Elizabeth tensed her face in worry. "Are we going to end up in the middle of the Atlantic?" she asked.

Lulu turned the first dragon's dial to 50. "Fear not a life pelagic!" she recited dramatically.

Next, Lulu turned the knob in the other dragon's mouth to 40. "The instructions said that the distance will be doubled. If the calculations are correct, Wonderland

---

[15] The difference between 1:35 and 4:15 is 2 hours and 40 minutes. Converted to hours, this is 2 hours and (40 min. ÷ 60 min. in an hour) or approximately 2.6666 hours. Multiply by 15 to get a longitude of 40.

should be close to England." Lulu decided that this made perfect sense, after all, since Lewis Carroll was a British author.

With a final verification of each dragon's dial, Lulu threw herself onto the bed. "And now… we fall asleep."

"Not again," Elizabeth groaned. Lulu turned off the flashlight, and the girls lay back awkwardly on the rough bedspread and stared into the darkness.

"Kipling has a poem about Fifty North and Forty West," said Lulu after a minute of silence.

"I know," said Elizabeth with another groan. Lulu took this as an invitation to recite (even though she was of the general opinion that reciting fine poetry requires no invitation):

> *"When the cabin port-holes are dark and green*
> *Because of the seas outside;*
> *When the ship goes wop (with a wiggle between)*
> *And the steward falls into the soup-tureen,*
> *And the trunks begin to slide;*
> *When Nursey lies on the floor in a heap,*
> *And Mummy tells you to let her sleep,*
> *And you aren't waked or washed or dressed,*
> *Why, then you will know (if you haven't guessed)*
> *You're 'Fifty North and Forty West!"*[16]

"And that's magic!" she whispered.

"That's nauseating… literally," said Elizabeth, already feeling queasy. "We're lying on a ship-bed reciting poetry about storms at sea. Do you have a feeling that

---

[16] Rudyard Kipling, "*Just So Stories for Little Children*", 1902

33

this challenge is over the heads of two girls from land-locked Colorado?"

Lulu nodded gravely. "We've never seen the ocean." (Aside from Wonderland, this trip to Quebec was the farthest they'd been away from home).

Elizabeth felt a surge of panic. She grabbed her sister's hand. "I don't even know how the heather looks," she whispered.[17]

Play Along: Calculate your longitude using a sundial, and your latitude using a home-made astrolabe and the North Star. Join the fun on page 201.

---

[17] She is referring to Emily Dickinson's poem "*I never saw a moor*", 1890

4. Fifty North, Forty West

Trrrrr. A telephone was ringing next to Lulu's ear. She tried to open her eyes, to sit up, to call out. But regardless of how hard Lulu willed these actions, she found herself unable to perform any of them. As Lulu quickly discovered, all the voluntary muscles in her body were still in a state of sleep-induced paralysis, comfortably unaware of her intentions. She was resigned to the role of an observer - a silent, blind observer.

This nightmare would terrify adults reading my book; The thought of being trapped in one's body would give any grown-up chills, Lulu decided. *It's the adults, not the children, who fear circumstances they cannot immediately control.* Lulu embraced new experiences, and in her current state of being — and not being - she did not feel fear, but curiosity.

Trrrrr. The phone continued to ring. "Fifty North and Forty West. Please hold," said a weary male voice. There was a mechanical click followed by a deep sigh from the phone-answerer. Lulu felt warm air rush across her face. The breeze was gritty, maybe even salty, but not unpleasant. *Like an intimation of summer.*

"Execrable inquiries," huffed the voice. "At this rate, I shall have no time for my figures." There was a great inhale as though the speaker was readying another sigh. Before he could let it out, the phone rang again.

"You've reached Magic. Please hold," he said in the same tired tone. Click. Click. "Petrus here. How can I help you? No, no, no. The Bermuda Triangle is at extension 25N, 71W." Click. Click. "Thank you for holding. What? No, sir. We cannot be held responsible. That has never been the policy. Good day." Click. Another windy, salt-tinged sigh swept over Lulu's face. "Now back to the figures."

Lulu heard the familiar sounds of scribbling – a pencil's soft scratching, a finger's anxious tapping, and the crinkling and shuffling of paper. She strained her ears, listening for more sounds. There were none. Aside from the scribbling, the room – wherever she was – was entirely devoid of ambient noise. When the scribbling stopped, there was only silence.

Had Lulu been in control of her muscles, she would have jumped when the voice suddenly yelled out. "We've got a live one, Captain!"

"Bring 'er in, Matey. By all means, bring 'er in," replied the same voice.

Lulu felt a drop of water hit each closed eyelid. The drops slid down her cheeks like tears and lingered on her lips. *Salty.* Lulu's eyes popped open. A grinning face was looking down at her. It belonged to a short bald man who took a step back and bowed.

"Welcome to Magic, the nautical navigation outpost at Fifty North, Forty West, and two thousand leagues below," said the man. He took a step back and swept his arms in a grand gesture.

Lulu discovered that she had regained control of her muscles from the neck up. Pressing her chin to her chest, Lulu was shocked to find the ship-bed jutting out from a wall. Whether the headboard dragons were in the room, Lulu could not tell, as she was unable to see behind her. What she did see in her field of vision was a rather disturbing image of the wall closed in around her shins (as if the lower portion of her legs, along with the bottom of the bed, had never existed). Lulu took a deep breath.

A circular window cut through the wall in front of her and beyond it was nothing but blackness- the darkest blackness imaginable. Lulu pictured two little white feet floating like jellyfish in the dark abyss. A green light flashed past the porthole. Lulu blinked. She squinted at the round window, then blinked again, but all she saw now was the blackness.

Lulu turned her head away. First, she looked to the right and found Elizabeth asleep beside her in a fetal position (so that no parts of her body had been cut off by the wall). Then, she turned her head to the left to get a view of the rest of the outpost, which appeared no larger than a closet. The bed (or, rather, whatever portion of the bed had successfully pushed into the room) took up most of the room.

The small, bald man was pressed up against another wall. He was grinning from ear to ear. The man held a medicine dropper in one hand and a piece of paper in the other. Behind him, the wall was covered with paper scraps. Lulu saw that each scrap was filled, top to bottom, with neat rows of numbers. Again, she felt the warm, salty wind.

The strange man, who was watching Lulu eagerly, cleared his throat. "Let me introduce myself," he said, throwing back his shoulders and raising his chin. "I am the magistrate of Magic, the guardian of portals and portholes, the master commander of wind and whimsy, official procurer of ships, and the constable of calculations and occasional chaos." He was counting the titles on his plump fingers. "You may call me Petrus," the man concluded with another bow.

"Pleased to meet you, sir," said Lulu politely. "But I think there's been some mistake. Magic was supposed to be the half-way point…"

Petrus's shoulders drooped and his smile faded. "It's true," he said glumly. "Magic is an in-between place. The Kraken tells of in-between places where people linger for days, weeks, months, years. Many of them have mistaken the in-between place for an actual destination. Take the tiny island of Arithmetic, for example. Most of its occupants believe they've reached Mathematics. How mistaken they are!"

Petrus looked longingly at the dark porthole. "At my outpost, visitors are rare and the long-term lodgers nonexistent. I do run a tight ship, after all." Petrus perked up slightly. "Still, the life pelagic can be a lonely one, especially after Nimue…" The melancholy-tinged words hung in the air like morning fog.

"My good friend the Kraken makes an appearance every other Thursday, at least," continued Petrus after a very long pause. "Good old Kraken does my bidding- collects misguided ships and such."

Lulu listened politely, afraid to interrupt the soliloquy. *A willing ear is the best gift to give someone who is unaccustomed to company*, she reasoned.

"Enough rambling, you scurvy dog!" Petrus suddenly yelled at himself.

"Captain's right. You're on a schedule, my dear girl. You're going SOMEWHERE. You're not a lingerer, that's for sure. I'll be sending you off immediately."

Petrus approached Lulu with a small medicine dropper. "This won't hurt," he assured her as he held the dropper's tip above her face.

"Wait!" Lulu yelled out. She wasn't entirely convinced that she was not a lingerer after all. "Before I leave, will you tell me a little about the work you do here? I'm very interested." The words came out, and Lulu was not sure if they were the product of curiosity, pity, or procrastination.

"I'm so very glad you've asked," said Petrus, beaming. "It would violate protocol to keep you here too long, but there is no harm in showing you one of my inventions." He looked to the left and the right of him before shoving a piece of graph paper in front of Lulu's face. "Don't tell the Captain," Petrus whispered loudly. Lulu nodded to indicate that she (mostly) understood.

"The Constable of Calculations must be adroit with arithmetic," Petrus continued. "The currents of both air and water depend on my figures. It's a critical job and a busy one. But when the seas are calm, there are fewer numbers to run. In those long, lonely hours, I turn into an inventor – a magician of sorts, if you will. And this-" Petrus waved the paper close to Lulu's face, "-this is my finest invention yet! To the nascent navigator, this may appear a pedestrian parabola, a commonplace curve. But I ask you to look deeper, to truly behold the brilliance of my magnificent multiplication machine."

"How does it work?" Lulu asked.

"Let's say you want to multiply four and three," Petrus began.

"Twelve!" Lulu answered.

"No, no, no. Let's say you don't know how to multiply four and three."

"O.K."

"That's where the magnificent multiplication machine shines. You find the square of four here," Petrus said, putting a little dot at coordinate (4, 16). "And the square of three here." He made another dot at (-3, 9). "Draw a line between the two points, and... voila! Your line intercepts the Y-axis at precisely twelve. The magnificent multiplication machine has found your answer!" Petrus grinned. "And that's magic," he added in a whisper.

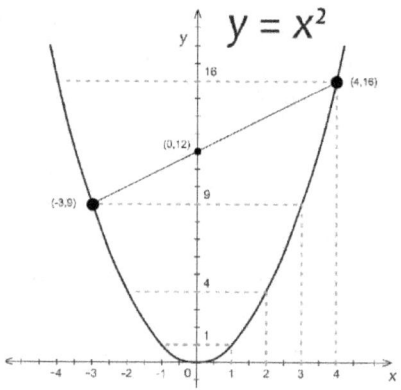

$$y = x^2$$

"That's quite clever," said Lulu with genuine approval, "but wouldn't it be faster to-"

"Shh..." Petrus interrupted. "I think I hear the Captain coming. I must send you off at once," he said, looking suddenly very nervous. "You will show my invention to Q, won't you?" he shoved the paper into Lulu's pocket and held the dropper above her eyes.

Before Lulu had time to ask who Q might be, the salty drops hit their mark. "A tender song from an old friend to see you off," Petrus whispered. A distant melody played in Lulu's ear (*was it Greensleeves?*). And just like that, Lulu was no longer a lingerer.

Play Along: From Magic, the 'distance will double.' At what coordinates will the ship-bed end up? Does Petrus's 'invention' actually work? Answer these riddles and more in this chapter's "Play Along" section on page 208.

5. Lily Maidens

Magic was only one of many in-between places for Lulu as she transitioned from dream to dream (never lingering longer than necessary). The remote ocean outpost and its eccentric caretaker were forgotten in the journey, tucked away in some grand vault of the subconscious.

But at Fifty North, Forty West, and two-thousand leagues below, the music persisted as a short, bald man sang of lands he had only seen in dreams and revived the lost memory of a dear, dear friend...

Lulu awoke with a start and immediately opened her notebook. Shining the flashlight on the page, she wrote:

> "*Sepals, petals, stinging nettle,*
> *Stamens, pistil, downy thistle,*
> *From whence the defense of tenderness?*"

This last dream had already turned to glitter, and the captured words were not enough for reconstitution. Everything was pitch black. Lulu could hear Elizabeth breathing heavily beside her. In the darkness, she could only picture the look on her sister's face. It would not be a peaceful one. Lulu placed a hand on her sister's chest, and Elizabeth sat straight up. "They were everywhere," she murmured.

A door creaked and the room flooded with light. Lulu and Elizabeth covered their eyes, blinded.

"How many petals has the lily?" came a voice. The voice was feminine, gentle but firm.

The girls sat for a moment in stunned silence.

"Sepals, petals, stinging nettle-" Lulu finally blurted out, fighting to open her eyes against the glare.

"How many petals has the lily?" the voice repeated.

"It is imperative that you answer," added a second voice.

"The categorical imperative, the whole foundation of how we live," a third voice chimed in. This one recited the words slowly and lyrically like a song.

The girls' eyes began to adjust to the bright light. Three women were standing in front of their bed. They wore identical floor-length dresses of flowing white fabric. A flower emblem was embroidered on each dress. Lulu, always fashion-conscious, noticed these emblems at once. Delicate golden thread outlined the shape of a lily upon each woman's heart.

Six petals! Lulu opened her mouth to answer but suddenly felt a forceful squeeze on her hand. "It's not what you think," Elizabeth whispered in her ear. "Three are petals, and three are sepals."

Sepals, petals, stinging nettle, thought Lulu.

43

"Three petals has the lily," said Elizabeth in an even, confident voice.

The woman who had posed the question smiled. "You have passed the first test," she said.

"Is this a botany exam?" Lulu asked before Elizabeth delivered a sharp jab to her ribs.

"In a way," the first woman answered.

"A symbolic way," added the second.

"*Through the Dark Sod - as Education -The Lily passes sure - Feels her white foot - no trepidation - Her faith - no fear -*"[18] recited the third.

Elizabeth gasped. "You were quoting-"

She did not have time to finish her thought before being interrupted by the first woman.

"The test was just a formality. Now, the time for proper introductions is at hand," she announced. "I am Lady Elfinhart, and it is with utmost pleasure that I present my sisters, Lady Lynette and Lady Olwyn," she gestured to the two women beside her, both of whom curtseyed. "We are the Lily Maidens of the House of Orkney and the Sisterhood of the Traveling Beds, defenders of honor and the three-petaled way. Welcome to Camelot!"

Now it was Lulu's turn to gasp. Lady Elfinhart waved her arms theatrically. Lulu and Elizabeth had been so focused on the Ladies, that they had taken no notice of their new surroundings. The ship-bed was sitting on a polished floor. Heavy velvet curtains hung around it.

The light was pouring in from above. A stained-glass window depicted the three Ladies standing side by side. Their angelic faces were raised toward the heavens.

---

[18] Emily Dickinson, "*Through the Dark Sod - as Education*"

Each Lady held one hand gracefully over her heart while the other hand clutched a perfect lily.

Lulu didn't know whether she should feel deeply impacted or burst out laughing. When she returned her gaze down from the stained-glass window, she found the ladies posing, still as statues, mirroring the looks in the glass to perfection. Their hands were on their hearts, and their eyes turned up toward the light. Only the flowers were missing from their hands. Lulu couldn't ignore the comedic irony of the scene. She was unaware that her mouth had fallen agape until she felt Elizabeth's hand on her chin. Lulu let reality sink in, and excitement soon took the place of disbelief.

"Are we really in Camelot? In the time of King Arthur and Queen Guinevere?" Lulu asked with a dreamy sigh. Elizabeth noted that the Ladies exchanged disapproving looks at the mention of Guinevere.

"Both correct and incorrect," replied Lady Elfinhart, swinging her dark shoulder-length hair aside.

"And also partially correct, which is an altogether different matter," noted Lady Lynette. She had golden hair, and Lulu thought that this Lady wore a more mischievous look than the other two.

Lady Elfinhart took a deep breath as though annoyed. Her even smile returned, however, as she resumed the explanation. "Some of the legends told about our realm contain accurate elements. Others reveal more about their authors than our reality."

Lady Lynette quickly chimed in. "It is now After Arthur, and women in A.A. times are different from what you might expect. We're modern ladies!" She seemed pleased to be forming her own sentences rather than echoing Lady Elfinhart's.

"That explains why they're not speaking Brythonic," Elizabeth whispered to her sister. This was a good thing, too, as the two girls would not have understood the language of the Celtic Britons.

Lulu was bursting with questions. She suddenly remembered her manners. "Pleased to meet you," she said, curtsying awkwardly. "I'm Lulu, and this is - "

"I know who you are, Lulu and Elizabeth Lovelace" interrupted Lady Elfinhart, "and the time for further introductions is at hand no longer. Mrs. Magpie has spoken very highly of you both. She sent you here to learn the three-petaled way of the Lily Maidens, but I believe you can assist us as well."

"Lily Maidens?" Lulu blurted out, her manners quickly forgotten. "Isn't that like saying you're a bunch of pansies? Weak!" She dodged another jab from Elizabeth's elbow and wondered if she would be thrown in a dungeon for speaking so blatantly to Ladies of the Court.

The smile did not fade from Lady Elfinhart's face. "Never underestimate the power of a flower. I'm sure pansies have their own redeeming qualities, but the lily represents the values we hold dear. She turned to Lady Olwyn, who began reciting on cue:

> *The lily sprouts a single leaf-*
> *The single truth we must defend.*
> *For honor's code is a belief*
> *We must preserve until the end.*
>
> *Three sepals guard a bud so pure*
> *With patience and integrity*
> *For growing knowledge must endure*
> *With courage and sincerity*

A three-petaled lily sheds its sheath,
Raises its head then gently bends,
For honor's code is a belief
We must preserve until the end."

This time the Ladies placed three fingers over their hearts and held these poses long enough to give Lulu and Elizabeth time to exchange glances and shrug. Lady Elfinhart opened her mouth to elaborate, but as she turned toward the bed, her face grew pale. "It's already happening to this one!" she yelled out. "We must check the others."

Lady Elfinhart pulled on a thick cord, and the velvet drapes fell with a thud, revealing an enormous chamber full of beds of varying sizes and shapes. The three Ladies scattered, fluttering from bed to bed like winter birds upon a field of wild rye. They ran around frantically, searching, their soft-souled shoes quietly tapping on the floor and their skirts rustling beneath them.

Lulu and Elizabeth watched as each Lady would arrive at a bed, let out a little gasp, and continue running to the next. Lulu wondered why they had spent so much time on riddles and introductions if this matter was of such importance.

"They've all faded," yelled out Lady Lynette.

Suddenly remembering the girls' presence, she turned to Lulu and Elizabeth and called out, "Help us find a bed whose jewel hasn't dulled." Lulu spotted what appeared to be a jewel on the headboard of a small pink bed. She dashed toward it, but upon reaching her target found that the gem was no longer glimmering but dull as a plain gray stone. Lulu let out a gasp. The sense of urgency was contagious. Elizabeth followed her sister's lead and headed for a medium-sized wooden bed that had been painted white. The gem on its headboard- a diamond- blinked twice and darkened before her eyes.

From bed to bed, Ladies and sisters dashed. Just as Lulu paused to catch her breath, she spotted a green glow out of the corner of her eye. It took the shape of a mouse. The mouse was perched at the foot of a bed that was covered with an elaborate quilt full of flowers and fruit. Lulu ran toward it, and the mouse vanished. The quilt was soft as down, and the mattress felt as if it were filled with water. Lulu swam across the quilt to examine the center of the headboard. An ornate flower was carved into the wood. It held an emerald in its center which was shining brightly. "I found one!" she yelled out. "This emerald is still glowing!".

Lady Elfinhart and Lady Olwyn soon dove onto the quilt, which rose and sank, heaving in great waves under the sudden disturbance. Lulu reached for the emerald and pricked her finger on the flower decoration. *From whence the defense of tenderness?* She observed as a drop of blood fell from her finger and landed upon the quilt, darkening an embroidered rose.

Lady Lynette rushed to join Lulu and the rest of the Ladies. Elizabeth was right at her heels when Lynette suddenly stopped next to a small bed. "Here's another!" she yelled.

"Take Elizabeth with you," Lady Elfinhart instructed. "With two beds, our chances of reaching Victoria will double." Elizabeth looked at the bed. It was a simple, sensible bed - one she would consider sleeping in if she was back home. Elizabeth focused on the headboard. The stone in its center was as dull as any of the rest.

Lady Lynette was studying her face. "We're too late on this one - it's faded," she called out. Without waiting for Lady Elfinhart to respond, Lady Lynette grabbed Elizabeth by the hand and pulled her toward the doorway. "We are going to take my horse," she yelled back.

Lulu glanced over her shoulder as Lady Lynette and Elizabeth disappeared. She stuck the wounded finger in her mouth and turned back to Lady Elfinhart, who was pointing to a sheet of paper next to the flower on the headboard.

"Read the instructions," Lady Elfinhart demanded. "Each one is different."

Lulu suddenly took notice of the paper. This note had been written by the same hand as the note in the Mouse box. Lulu read the instructions aloud:

Bed #229. Default: 51.02, 2.53

"Quickly, quickly, without hesitation,
Enter the sepals for your destination.
Remember that only the roses will do,
And enter your answer using base two."

"The quilt is a map," said Lady Elfinhart. "Look for Dozmary Pool. That's where we'll find Victoria."

Reading the labels on the quilt was like tracking a single salmon in a moving stream. The more frantic the three searched, the more the quilt moved. Lulu took the approach of the grizzly bear, sending forth a broad paw (which in reality was only a feeble hand), and then reading any labels in its vicinity.

It was Lady Olwyn who was successful. She held her finger firm against the current. "Dozmary Pool," the label read. Lady Elfinhart and Lulu froze, and the quilt's movement slowed. They moved in slow motion and smoothed out the fabric until it lay almost flat.

A blue line outlined the boundary of the Dozmary Pool section. Inside the borders, lay tiny embroidered roses and fruits. The details on these were so fine that Lulu decided that the needlework could only have been accomplished by the fingers of fairies. "How do I count the sepals?" Lulu asked, turning to Lady Elfinhart.

An impatient sigh rose from the Lady. "Carl gave us the impression that all modern men and women were well versed in botanical terms."

"Carl Linnaeus?" ventured Lulu.

"Who else?" scoffed Lady Elfinhart. "You must count by fives since roses have five sepals each. Remember that time is of the essence."

Lulu began counting. "Five, ten, fifteen, twenty-" She paused with her finger on a lovely round plum. "Is this a... rose?" She felt silly asking the question and was waiting for another rebuke from Lady Elfinhart.

"Of course," Lady Elfinhart replied. Lulu was thoroughly baffled. She wrinkled her nose.

"Why, oh why, do you suppose, Lulu shrinks at the sight of a rose?" asked Lady Olwyn.

Later, as Lulu would replay these events in her head, she would form a fantastic retort:

> *"Who wouldn't become decidedly glum,*
> *At the sight of a rose dressed up as a plum?"*

But, as the best responses are always formed too late, Lulu only smiled and resumed counting by fives. She stopped again on an apple, then on a peach, and finally a cherry. At each of these fruits, Lulu looked at Lady Elfinhart who gave a nod of approval. Soon Lulu was counting everything, never taking her eyes of the quilt.

50

Pears, peaches, strawberries and even almonds were apparently in the rose family. This was news to Lulu, but, as Lady Elfinhart had said, "time was of the essence."

Lulu counted "-seventy-five, eighty, eighty-five, ninety." Her finger rested on the last item - a fig. Lady Elfinhart and Lady Olwyn both gasped loudly. "Eighty-five, then," Lulu corrected herself.

Lulu read the instructions again "*Enter your answer using base two.*" The Ladies exchanged looks. "Using base two," Lulu repeated.

Lady Olwyn answered:

"Ipomoea greets the glorious morn,
Unrolling beauty dew has adorned.
Mourn not a mind that sits forlorn,
For it, too, in time, shall be reborn."

"Ipomoea is a morning glory," explained Lady Elfinhart curtly.

"And how is this related to 'base two'?" Lulu asked, frustration building up in her voice. For Ladies in a hurry, they were moving at a laughable pace.

"With every day, comes a new beginning," Lady Elfinhart replied.

"How is this related to 'base two'?" repeated Lulu. This time, she was yelling.

An awkward silence followed. "We've never personally activated this bed," Lady Elfinhart finally admitted, her painted smile turning to a sheepish grin. "We were hoping you'd know what is meant by *base two.*"

Lulu almost laughed out loud. *I'll have to remember to use that rhyme next time I don't know an answer.* She turned her attention back to the headboard. Next to the large rose decoration which had pricked her finger, was a row of small buttons. She pushed one, and little metallic flower petals sprung up around it. She pushed the button again, and the petals retracted.

"Two states - open and closed," Lulu mumbled to herself. She understood immediately. Up until now, she had felt inadequate, as botany was not her field. But this was math and computer science. Lulu was in her element. "Base two means binary!" she said. The Ladies wore baffled expressions.

Lulu grabbed her notebook and wrote: 64, 32, 16, 8, 4, 2, 1. She had to combine these values to make the number eighty-five[19]. Lulu wrote 1 beneath the 64. Subtracting 64 from 85 left 21, so she wrote 0 beneath the 32. Continuing in this fashion, Lulu soon had the binary number 1010101 (since 85 = 64+16+4+1). She pushed the buttons on the headboard, guessing that an open flower was a "1" and a closed flower a "0". "Done!"

"Well done!" said Lady Elfinhart.

"Well done and hard won,
Ipomoea, disrobed,
Has her day in the sun!"

Lady Olwyn's voice was sweet and gentle. The two Ladies immediately lay down on the bed and motioned for Lulu to join them.

As Lulu tried to fall asleep, her mind raced. She made a mental note to include this latest puzzle in her novel. Would her readers require more explanation, or had she provided enough of a clue for them to figure out Base Two on their own? She might suggest that her readers count the unique combinations that can be

---

[19] Learn binary counting in this chapter's "Play Along" section on page 216.

formed with two on-off switches and then add another switch and another..."

"Look for patterns," Lulu mumbled. There was more to think about, but the chasm between right and left hemispheres was widening rapidly. Soon logic was swept into a world of whimsy, where numbers and flowers dance - elegantly, infinitely, on delicate feet.

Play Along: In this chapter's activities, you'll familiarize yourself with the British folklore of King Arthur and find out why three is a magic number. You will also learn to identify the parts of a flower, explore the members of the rose family, and learn binary counting (base 2). Join the fun on page 212.

6. The Fish of Desire

Forty moments fill a sun's hour,
But what's more ephemeral than a flower?

For Lulu, waking up with a rhyme on her tongue was not an unusual occurrence (*doesn't poetry provide the best soundtrack to life?*). The strange rhymes that filled her head when she slept upon the traveling beds were different, however. The unfamiliar tunes had a heaviness to them that Lulu couldn't quite place. "What's more ephemeral than a flower?" she asked aloud.

"A dream," answered Lady Elfinhart, startling Lulu into full consciousness. "Fleeting, but prophetic, it appears. We must continue by foot."

"Where's Victoria?" asked a confused Lulu.

Lady Elfinhart pointed to the gem on the headstone of the bed in response. The center of the ornamental flower had turned from emerald green to midnight black. Lady Olwyn sat up on the bed and smoothed down her dark curls.

The bed had landed in the middle of a clearing, surrounded by lumbering pines. Elongated purple shadows, like witches' fingers, reached for the delicate flowered quilt.

Lady Elfinhart looked up at the sky. "Forty moments[20] and daylight will fade. We do not want to be here when dusk descends into night." Lulu agreed without question.

"We move to the west," said Lady Elfinhart, pointing toward distant mountains and the setting sun. In front of them stood the thickest part of the pine forest. Lady Elfinhart gathered up her skirts and began marching. Lady Olwyn followed, copying her manner of walking like a shadow. Lulu was unaccustomed to being a follower rather than a leader. However, reasoning that she didn't know the way, Lulu soon found her place in line behind the two Ladies and began trudging through the forest.

Birds were perched on every tree, watching, listening, as the parade of Ladies marched by. A spotted nutcracker sat on a low branch and cocked its head as it studied the procession. Lulu thought longingly of Mrs. Magpie. *Is this a lesson in obedience, maybe even humility?* she wondered. Still, Lulu couldn't shake the feeling of uneasiness at having to follow blindly and silently.

As they walked, Lady Olwyn began reciting the beginning of a poem that Lulu recognized as the work of Henry Wadsworth Longfellow:

> "*This is the forest primeval.*
> *The murmuring pines and the hemlocks,*
> *Bearded with moss, and in garments green,*
> *indistinct in the twilight,*
> *Stand like Druids of eld,*
> *with voices sad and prophetic,*
> *Stand like harpers hoar,*

[20] A moment is a medieval unit of time based on the movement of a shadow on a sundial.

with beards that rest on their bosoms.
Loud from its rocky caverns,
the deep-voiced neighboring ocean
Speaks, and in accents disconsolate
answers the wail of the forest."[21]

The poem could not have described the forest more accurately. "We do not want to be here at night," repeated Lady Elfinhart.

Lulu ran up to Lady Elfinhart's side. "I'm a smart girl," she began.

"I do not doubt it," said Lady Elfinhart as she continued walking. Her face was soft, and the corners of her mouth turned up.

"-and smart girls know when to follow and when to lead," continued Lulu. "But most importantly, they ask questions." Lulu felt a sense of relief as the words flowed out. "I have plenty of questions, and the more I know, the more helpful I can be."

Lady Elfinhart smiled. "Without a thirst for knowledge, we would not be here. Asking questions is part of the three-petaled way of the Lily Maidens and the Sisterhood of the Traveling Beds."

Lulu was relieved.

"But remember, Lulu," Lady Elfinhart continued. "Knowledge is never handed to you. You must take the initiative to seek it."

Lulu felt foolish for simply falling behind in line and assuming this was what was expected of her.

"Question number one: explain the beds," Lulu demanded, hoping she didn't sound impudent.

---

[21] Henry Wadsworth Longfellow "*Evangeline: A Tale of Arcadie*", 1893

"That was not a question, but an imperative," corrected Lady Elfinhart.

Lulu was surprised when Lady Olwyn was the one who answered her query:

> *"We travel the world through legends and time,*
> *For often they will intertwine.*
> *We sail upon beds and poems and myths,*
> *The draw of knowledge is hard to resist."*

"An elegant introduction," said Lady Elfinhart, "but we must start at the beginning. And, as all stories involving magic in Camelot do, this one begins with the Fairies." Lulu couldn't mask her delight.

"Your reaction is typical among those unfamiliar with the true nature of Fairies. I remember meeting a young girl named Enid Blyton who wore an expression very similar to yours at the mention of Fairy folk. She covered her ears and hummed loudly when I warned her of the dangers."

Lulu had read Blyton's *"Book of Fairies"* (as well as her *"Magic Faraway Tree"* trilogy). In the stories, brownies, gnomes, and elves would get into mischief, and the beautiful, good-hearted fairies always saved the day.

"Listen carefully," continued Lady Elfinhart. "Fairies are the keepers of the supernatural realm and can grant mortal desires. They can be munificent, that is true. But take heed. Their generosity always comes at a price. There was a man named William Butler Yeats who understood this all too well. He was the first seed, you see."

Lulu gasped again. "How?"

"Are you familiar with Yeat's poem, *"The Song of the Wandering Aengus"*?

Lulu nodded.

"That poem makes up only half the tale. The full story begins with a different poem - a poem that came to young William in a dream." She turned to face Lady Olwyn, who recited in a gentle voice:

"Burning fire,
Hazel, Rowan,
The fish of desire,
The price of knowing."

Just like my own rhyming dreams, thought Lulu. Lady Elfinhart raised a hand as though Lulu's thought was interrupting the story. "The time for narratives is upon us," she said and continued walking at a brisk pace. Lulu and Lady Olwyn stuck close to her side. As they walked on, Lady Elfinhart told them the tale.

William sat up in his bed. Sweat dripped from his cotton tunic. "Another dream," he mumbled. The dreams had been haunting William for a week now - always rhyming, always sung by the same alluring voice. "The fish of desire" she had said, lingering on the final "r" so it sounded like tumbling water. Suddenly William felt a burning sensation deep in his belly. It first took the form of a passionate hunger. The small spark ignited the moment William had acknowledged its presence. The fire's long fingers quickly moved upward to his chest and across his arms. William's head was consumed with intense heat. He wiped the perspiration from his forehead with a damp sleeve. The coals in the fireplace still held an orange glow, but the fire itself had long died out. *Did he have a fever?*

William unbolted the door and stepped into the night air. His right hand instinctively moved to his pocket where it wrapped around his knife. William began walking.

Walking at night was unusual for William as he was a man who believed in Fairies. He had long ago shrugged off the false friends who had scoffed at his belief in the supernatural. Those individuals blamed his "madness" on the Irish folklore he was always reading. But William believed in Fairies, not because of old stories, but because he had seen things with his own eyes - strange lights hanging over the lake, trees that whispered. Only yesterday, an eerie green glow had appeared just outside his window. *The woods were alive!*

William would always shut and bolt the door at twilight, retreating into the safety of his home and his books. *But how everything changes with the burn of desire!* Walking into the woods, William snapped off a thin branch of hazel. He whittled at this stick with his knife as he made his way to the lake. *Was he fashioning a weapon?* A Rowan tree caught his eye. Her fruit sparkled in the moonlight, glowing as red as the embers in his belly. William stopped to scoop up handfuls of the glistening fruit to fill his pockets.

The lake was a sea of darkness. *How different it looked at night!* The swans (he had counted fifty-nine of them earlier that day) were nowhere to be seen. William stirred the water with his hazel branch. He reached into his pocket, pulled out a rowan fruit, and tossed it into the lake. The small sphere glowed as it hit the water and immediately disappeared. William threw in a second fruit, which also flashed and faded like the light of a firefly.

Hypnotized, burning, William threw in another and another and another. He counted nineteen fruits, each glowing brighter than the last and vanishing more rapidly. William's hand reached the bottom of his pocket. He rolled the last gift of the rowan between his fingers. William hesitated. Something splashed near the spot he had thrown the berries. *Was it a silver fin he saw?*

William pulled out the last fruit and was surprised to find that thread from his pocket had wrapped around his hand. A needle twirled from the end of the thread like a compass. William knew what he must do. The fire commanded it.

The fruit was soon secured to the hazel stick, and William cast the makeshift fishing line into the water. He did not realize it at the time (nor would he ever quite understand), but this particular act released an ancient magic - a Fairy magic - into the lake at Coole Park. All the elements were present - burning fire, hazel, rowan. Unbeknownst to William, he had already made a wish - this was the burning fire. The hazel, in turn, unlocked the magic within the rowan fruit and changed it into a gem - a perfect diamond.

Blinded with monomania (or perhaps it was the fever), William was oblivious to the fruit's transformation. All he could see was the fish at the end of his line.

> *"I went out to the hazel wood,*
> *Because a fire was in my head,*
> *And cut and peeled a hazel wand,*
> *And hooked a berry to a thread;*
> *And when white moths were on the wing,*
> *And moth-like stars were flickering out,*
> *I dropped the berry in a stream*
> *And caught a little silver trout."*[22]

---

[22] William Butler Yeats "*The Song of Wandering Aengus*", 1899

Raging with hunger, William ran straight to his cabin and flung open the door. The fish dropped to the floor with the thread still hanging from its mouth as William turned his attention to the fire. It was soon crackling pleasantly. That's when he heard a voice calling his name. "William, William Butler Yeats." It was not just any voice that was speaking to William - it was the voice from his dreams!

William spun around to see a vision of a beautiful woman. Her gown was as silver and shimmering as fish scales with sleeves of emerald that reminded William of both sea and pasture. Her hair and lips were as red as the rowan fruit. William lowered his eyes from the glimmering girl. *The fish of desire!*

The vision walked toward him upon the smallest, palest, most delicate feet he had ever seen. "William," she sang. "William!" William bit his lower lip, and the taste of warm blood filled his mouth - reassurance that this was not a dream.

The woman - for she was no vision, but flesh and blood - lifting his chin gently. Their eyes met, and the fire crackled. A velvety green sleeve pressed against his mouth, and the woman leaned closer. Silky red hair brushed William's cheek, and the smell of apple blossom filled the air (*or was it the scent of lilies?*). He closed his eyes.

The woman planted a single soft kiss upon each eyelid. The sensation was that of a cool drink. William was no longer on fire.

When William opened his eyes, he saw only a flash of silver and emerald and crimson at the open door. The tune to "*Greensleeves*" rang in his head. William's soul beckoned him to run after the woman, but his body moved in slow-motion as if he were walking through water. When William finally reached the threshold, the glimmering girl had already disappeared into the night and his eyelids felt heavy, *so very heavy...*

> *"When I had laid it on the floor*
> *I went to blow the fire a-flame,*
> *But something rustled on the floor,*
> *And someone called me by my name:*
> *It had become a glimmering girl*
> *With apple blossom in her hair*
> *Who called me by my name and ran*
> *And faded through the brightening air."* [23]

William would see the woman again, but only in his dreams. Sometimes he would sleep for days at a time as she sang to him in that beautiful voice. *Was the woman Caer Ibormeith, the Celtic goddess that guarded the realm of sleep?*

William would return to Coole Park every fall. He tried summoning the magical fish again and again. But although he would occasionally catch a glimpse of a mysterious green light in the dark woods, neither girl nor fish ever materialized.

The glimmering girl became a vision that would haunt all of William's sleeping and waking moments. Every night, his love for her would deepen, but every morning he would wake up alone. The smell of apple blossoms (*or was it lilies?*) followed William as he wandered the Irish countryside. This was the price of knowing.

> *"Though I am old with wandering*
> *Through hollow lands and hilly lands,*
> *I will find out where she has gone,*
> *And kiss her lips and take her hands;*

---

[23] William Butler Yeats "*The Song of Wandering Aengus*", 1899

And walk among long dappled grass,
And pluck till time and times are done,
The silver apples of the moon,
The golden apples of the sun." [24]

Nineteen years later, as William Butler Yeats, now an esteemed poet, was counting the wild swans at Coole Park, he noticed a pair of swans that would not leave each other's sight. They swam through the lake apart from the other swans, their heads always touching. As they took to the sky, the pair flew in synchronized motion, always singing. Yeats was reminded of the legend of Aengus, the god of love.

Aengus, too, was visited in his dreams by a beautiful woman that he could not get out of his mind. The mere thought of the woman tormented him for a year before his desire culminated in a life-threatening illness.

Burning with fever, Aengus finally revealed the secret of the mysterious dream-woman to his family. Desperate to find a cure for their son's ailing heart, Aengus's parents had the entirety of Ireland searched for the girl. It would be two more years before Aengus, now barely clinging to life, received the word that the woman had been found. Her name was Caer. It would be longer, still, before she could be his, as Caer's father would not agree to even speak to Aengus's family about the matter.

A war ensued between the two kingdoms, and much blood was shed. Finally, the parents of the two lovers made peace. Caer's father revealed that he had not wanted to give up his daughter because she possessed

---

[24] William Butler Yeats "*The Song of Wandering Aengus*", 1899

great power that allows her to take human form for one year and the form of a swan the next.

Aengus met Caer on the shores of Loch Bél Dracon, The Lake of the Dragon's Mouth. One hundred and fifty swans floated like silent ships on its waters, but Caer was in her human form. Before giving Aengus her hand, she asked that he allow her to return to the water. Aengus gave his word, pledging his love and her freedom. Then, at last, the woman he had seen in his dreams for so long came to him. The moment they touched, both Aengus and Caer turned into swans.

Yeats held his breath as he watched the pair of wild swans at Coole engage in an enchanting dance upon the water. The melody of a familiar tune played in his head (*was it Greensleeves?*) William let out a sigh and thought of his own glimmering girl. *Maybe, just maybe, there was still hope...*

"And that's the first half of the story," said Lady Elfinhart.

"What happened to William? Did he ever find the woman?" cried Lulu. She was so absorbed in the tale that she hadn't noticed the sun vanishing behind the mountains. A full moon now took the sun's role as the Ladies' guide through the Forest Primeval.

Lady Elfinhart shook her head sadly. "Yeat's brush with Fairies left within him an insatiable longing, a pain in his heart that would never heal."

"The heart sometimes leads the mind astray. When you deal with Fairies, there's a price to pay," sang Lady Olwyn.

Lulu retrieved her notepad.

"I hope you're not planning to tell the world that Yeat's poem, "*The Song of the Wandering Aengus*", is autobiographical," said Lady Elfinhart, looking sharply at Lulu. "No one will believe you. Scholars, not dreamers make such determinations."

Lulu thought that at this moment Lady Elfinhart sounded a lot like a certain sister of hers. Thinking longingly of Elizabeth, Lulu scribbled a few words in the notepad and returned it to her pocket. *It was too dark to write properly, anyway.*

"Yeats role was minimal when you look at the big picture," Lady Elfinhart added. "Stirring sticks in the water, carelessly tossing fruit, burning with youthful passion - Yeat's greatest contribution was telling the tale through his poetry, nothing more. Nimue was the real hero of the story."

"Nimue?" asked Lulu. The name seemed oddly familiar, but Lulu could not recall where she had heard it.

"Nimue was The Lady of the Lake at the time. She had taken the form of a fish that night. When Yeats unwittingly unleashed the Fairy magic, Nimue swallowed the diamonds that formed. The last, still threaded on the hazel wand, caught her by surprise (*or was it fate?*).

You know what happened next from Yeat's account. As soon as Nimue had turned back to a maiden, she returned to the lake. With the diamonds still in her belly, Nimue emerged in Camelot, which you now know is neither place nor time nor legend, but rather some combination of the three."

Lulu nodded.

"The magic diamonds started it all," continued Lady Elfinhart. "Nimue brought them to the Lily Maidens. At the time, we were only a loose organization of women. Always loyal to our husbands, brothers, and

65

King, we aided Camelot through whispers and intimations. We knew the jewels were the key to knowledge, which, in turn, is power. Power is safer when entrusted to the gentler gender."

At this, Lulu nodded emphatically, and Lady Elfinhart's smile widened before she resumed the explanation. "Nimue was something of what you might call- in your time- an inventor. In her mountaintop laboratory, she devised a way to use the jewels to travel to other realms. When she met with the Lily Maidens to report her discoveries, she said that we would need vessels to aid us in our journeys. I suggested that we use beds. It was perfect. Arthur was still alive in those days. Knights saw all the action, and women were mostly confined to their bed chambers."

This was apparently a cue for Lady Olwyn to sing:

> "*How do Arthurian ladies*
> *Pass the time in their beds?*
> *While the men are crusading,*
> *They dream of adventures ahead.*
>
> *Dreams turn damsels to drifters,*
> *To sailors and scholars of time.*
> *Beds become frigates,*
> *Gems turn the spigots*
> *Of seas and stories sublime.*"

"The Sisterhood of the Traveling Beds has visited with renowned authors, inventors, and leaders from many places and time periods. Most see us in their dreams, while a select few are entrusted with our secret. We also jump into works of literature. All that we learn, we try to apply to our code of honor and toward the good of Camelot."

"But what's happening to the beds now?" asked Lulu.

"The gems come in pairs. One is embedded into a bed's headboard while the other lies at the bottom of Dozmary Pool. The water has special properties that give the gems energy and connect them to other realms. Victoria is the guardian of Dozmary Pool and the gems it holds. We need to find out what is happening to them."

"Victoria sounds like the name of a queen," said Lulu

"It *is* the name of a queen, of course. I thought you'd know that." scoffed Lady Elfinhart.

There was a long pause. "I still think Nimue was cruel to tease William like that," Lulu finally said.

"Nimue, The Lady of the Lake, was a Fairy herself. The circumstances were out of her control as she was under the command of the Queen."

"Victoria? Guinevere? " asked Lulu.

"Fairies don't abide by the laws of mortal monarchs. Nimue was in the service of the Fairy Queen. This is where the story continues."

Lulu listened intently as Lady Elfinhart continued the tale.

There were twenty gems in the beginning - enough to power ten vessels. Nimue brought all twenty to the Fairy Queen for safe keeping while she perfected the technology for the traveling beds. At the end of the first week, three beds were ready – a detail you mustn't forget as there is magic in the number three. Nimue came to the Fairy Queen to retrieve six jewels, leaving fourteen.

Another week passed, and the Lily Maidens procured three more beds. Nimue withdrew another six jewels, leaving the Fairy Queen to guard the remaining eight. During the third week (*that's magic*), Nimue had prepared the final four beds. Here is where the trouble began.

On her visit to the Fairy Queen's abode, the Queen was already in a foul mood. After retrieving the remaining eight jewels (leaving none), Nimue discussed her plans to entrust Victoria with one gem from each pair. For some reason, this information sent the Fairy Queen into a rage.

"I enchanted your jewels and produced two more for you!" she screamed. "There were fourteen jewels left after the first week, and eight after the second. That makes twenty-two. I created two new jewels for you. What have you done with them? I bestowed you with a precious gift, and you repay me with thievery!"

Nimue was mortified. There were twelve jewels in the bed chamber and eight in her hands. *How could she have lost two of the jewels?* Nimue fumbled for an explanation. She took two of the eight gems from her palm and held them out to the Fairy Queen. The Queen immediately snatched the treasures from Nimue's hand. She seemed appeased for the moment.

The mystery of the two missing jewels burned at Nimue. She searched every room in the castle, but never found them. An air of melancholy surrounded Nimue from that day on. Over the next year, she would spend most of her time in self-imposed isolation in her garden laboratory. There were whispers that she was shirking from her duties as the Lady of the Lake. But Nimue was on the brink of an incredible discovery.

One day, as the Lily Maidens were meeting, Nimue staggered into the bed chamber. Her face was as pale as the moon. Nimue began opening her apron but collapsed. A dozen jewels bounced and scattered across the floor. She had found a way of creating them in her lab!

As the Lilly Maidens rushed to Nimue's aid, a gust of wind filled the chamber. The Fairy Queen came tromping in. Thud-thud-thud went her feet (the Queen was always ashamed of her large feet and anyone who was foolish enough to utter a word about their size, would endure her wrath). She swept Nimue into her arms and disappeared. None of the Ladies tried to stop her, as Nimue was a fairy herself and thus the rightful property of the Queen. That's the last time we saw our friend.

Some say Nimue died of shame, punishing herself for the missing jewels, which she never recovered. Honor is everything to us Ladies. I, however, have my own suspicions.

Lady Elfinhart did not elaborate. "Thanks to the new jewels, our humble fleet of nine beds grew to fifteen, and our knowledge increased hundredfold," she concluded. "Does that answer your question?" The story was over.

Lulu nodded as she quite enjoyed the tale. They continued to walk silently through the dark woods. The history of the beds fascinated Lulu, but she couldn't stop thinking about The Lady of the Lake. "Nimue," the woods whispered. "Nimue."

Play Along: Explore the poetry of William Butler Yeats, grow sucrose crystals, and solve the missing gem puzzle in this chapter's "play along" section. Join the fun on page 220.

7. Golden Rowan

Lady Lynette grasped Elizabeth's hand and led her through dimly-lit corridors beneath the castle. Whenever Elizabeth opened her mouth to speak, Lynette would hold a finger over her lips to indicate silence. Elizabeth eventually concluded that this was for the better, as the dark passages had a strong smell - a foul blend of mildew and excrement.

Elizabeth was exceptionally sensitive to odors *We must be below the royal stables*, she told herself. *We're below the royal stables.* Elizabeth hoped that this was true. Having read a bit on the topic of medieval hygiene, she knew that they could very well be walking beneath the castle garderobe[25] – a worse alternative. Elizabeth didn't want to think about it. Instead, she looked down at her feet and focused on perfecting the art of minimal breathing as Lynette pulled her through the passages.

The corridor ended in a flight of stone steps and a door to the outside world- *light at the end of the tunnel!* Elizabeth let out a joyful breath as her nostrils met with fresh oxygen.

Green fields stretched in every direction and birdsong filled the air. Elizabeth was so relieved to see

---

[25] The castle toilet!

that they were indeed standing next to the stables that she almost let out a laugh.

Lynette disappeared. She returned shortly leading a white stallion from the stables.

Elizabeth came to her side and spoke at last. "Lady Lynette?"

"Just call me Lynette," the Lily Maiden replied, placing a delicate hand on the horse's mane.

"How far away does Victoria live?" Elizabeth asked. "That is where we are headed, right?"

"We'll get there eventually – after a short but crucial detour," Lynette responded as she busied herself with the horse's saddle.

"But Lady Elfinhart - "

"Lady Elfinhart is not in charge," said Lynette, a touch of hostility in her voice. "All three legs of a pedestal bear equal weight."

If they are the same length and on a flat surface, thought Elizabeth.

"Do you do everything your sister tells you?" Lynette asked.

Elizabeth was quiet.

Lynette led the horse past the stables. Attached to the side of the building was a pigeon loft with nine small cubbyholes. Eight of the nooks held a single cooing pigeon, but from the ninth pigeon hole, two feathery heads appeared.

"The pigeon hole principle!" said Elizabeth aloud.

Lynette looked at her for an explanation.

"If there are ten pigeons and nine holes, then two of the pigeons have to share."

"Is that not plain?" Lynette asked. She flicked a strand of hair off her face.

"It's an important math principle..." Elizabeth began, but her words suddenly seemed small, inconsequential.

Lynette grabbed a pigeon and stuffed it into the saddlebag. "Pigeons may be old-fashioned, but sometimes they're more reliable than Populus Post," she explained. "Now we must be off to Mendel's garden."

Soon Elizabeth, who was afraid of mice and wary of boats, found herself atop the horse, holding tightly to Lynette as they galloped through the countryside. If Elizabeth had not been frozen by fear, she could have glanced behind her and seen the grand castle of Camelot that she had only read about in legends. It was enough consolation to Elizabeth, however, to see the purple waves of the heather for the first time.

> *"I never saw a Moor —*
> *I never saw the Sea —*
> *Yet know I how the Heather looks*
> *And what a Billow be."*[26]

Above the sea of purple, Elizabeth spotted a green hill which held the white outline of a horse. "We are headed to Hengoen Hill," yelled Lynette.

Elizabeth, at once, was reminded of the *"Ballad of the White Horse"* by Chesterton.

> *"Before the gods that made the gods*
> *Had seen their sunrise pass,*
> *The White Horse of the White Horse Vale*
> *Was cut out of the grass.*
> *Before the gods that made the gods*
> *Had drunk at dawn their fill,*

---

[26] Emily Dickinson, *"I never saw a Moor"*

The White Horse of the White Horse Vale
Was hoary on the hill.
Age beyond age on British land,
Aeons on aeons gone,
Was peace and war in western hills,
And the White Horse looked on."[27]

Their own white horse stopped near a large linden tree. "We'll have to continue on foot from here," said Lynette as she tied up the horse. "But first, we must scour." She knelt and began pulling up weeds around a line of chalk stones. "Chesterton was one of the first poets we contacted with the jewels," Lynette explained.

"Guinevere was a Lilly Maiden back then- before she broke the vows of the three-petaled way." Lynette's face bent into a scowl. "Guinevere would whisper Chesterton's words as Arthur slept, and before we knew it, he was ordering his very own horse-on-the-hill - in memory of Hengoen, Arthur's beloved stallion."

Lady Lynette sped up her scouring. "It would be overrun with weeds if it was not regularly maintained by the Lily Maidens. Lady Elfinhart says we must do it in Arthur's memory and to remember all the brave knights who have shed blood for Camelot. I do it to remember that even honor's code must never be neglected. The human soul, too, requires maintenance."

"And Lady Olwyn?"

"She recites Chesterton, of course."

"And though skies alter and empires melt,
This word shall still be true:
If we would have the horse of old,

---

[27] G. K. Chesterton,*"The Ballad of the White Horse"*, 1911

Scour ye the horse anew."[28]

Lynette pulled up more weeds around the white rocks, and Elizabeth soon joined her. "Are you sure we have time for this?" she asked, as gently and politely as she could.

"You can always make the time to scour your horse," replied Lynette, pulling up another handful of weeds and tossing them into the wind. Elizabeth had to admit to herself that there was something peaceful, almost meditative, in the rhythmical work of scouring the horse on the hill.

As they worked, Lynette told Elizabeth the story of Nimue and how the gems and the traveling beds originated.

"The three-petaled way - what exactly is it?" Elizabeth asked.

Lynette did not look up. "That's actually a source of contention among the Lily Maidens," she said. "We all agree that we should hold ourselves accountable to the highest moral standards. The problem is that we cannot seem to come to a consensus on the three most important values to use in our motto. I believe that at the moment, they are 'honesty', 'courage', and 'perseverance'. Sometimes 'integrity' is thrown into the mix. Other times, 'respect' or 'loyalty' find a home among the three.

"Why not have more than three values?" Elizabeth asked.

Lynette stopped her work and looked Elizabeth in the eyes. "We're the Lily Maidens, protectors of the three-petaled way!" she said emphatically, straightening up and holding three fingers over her heart.

---

[28] G. K. Chesterton, "The Ballad of the White Horse", 1911

Elizabeth shrugged. *Symbolism loses its power when used in excess*, she decided.

Lynette wiped her hands on her white gown. "That will suffice," she said and began climbing an outcrop of rocks. She motioned for Elizabeth to follow. The cliff was steep and the terrain rugged, but it was not long before they had reached the top.

"Welcome to the Garden at Menalowan," said Lynette. "Nimue would always refer to it as Mendel's Garden." She smiled now for the first time.

Elizabeth looked around the lush hilltop garden. Fruit-laden tree-boughs framed the breathtaking view. Neat rows of rich brown soil stretched across the length of the garden. Wildflowers abounded at its perimeter – colorful and beautifully unkempt. In the corner of the garden, stood a wooden shed, painted white and surrounded by white lilies that stood tall in perfect rows. At that moment, the thought dawned upon Elizabeth that her companion resembled more of a wildflower herself than the neatly cultivated lily. She dared not mention this observation.

Lynette took Elizabeth by the hand and guided her to the shed. "This was Nimue's laboratory."

The shed held three rows of long benches lined with flasks, gardening tools, and an assortment of curious instruments.

"Nimue was working on a technique to create new gemstones so we could expand our fleet of beds."

"Was she successful?" asked Elizabeth.

"Yes, she was. But Nimue disappeared soon after revealing her discovery – before she could share her secrets with the Lily Maidens."

Lynette ran her fingers across the various articles that lined the workbenches. "I have a general idea of how the process works, although I've never actually tried it," she admitted. "But think of the possibilities! Creating new gems could not only solve our current problems but will also allow us to travel to unfamiliar and wonderful places we have yet to imagine!"

Elizabeth moved to the first bench on the left side of the shed. On this bench, sat a pitcher of water. Next to it was a machine of some sort. At its top was a large glass vessel filled with small glossy berries in red, yellow, and pink. Further down, the device sported a row of six metal knobs and a large dial. To Elizabeth, the contraption bore an uncanny resemblance to a gumball machine.

"Rowan fruit," said Lynette. "Nimue described this device to me once. I think I can figure out how it works." Lynette put a finger to her temple. "If I remember correctly, we need to get two berries of the same color."

"I think there are instructions," said Elizabeth, pointing to a piece of paper on the table.

"Modern paper," gasped Lynette. "This is Nimue's work, without a doubt." She picked up the paper and admired its crispy white gleam before reading the message aloud:

"*Golden Rowan of Menalowan revealed her wedding gown,*
Delicate pinks, yellows, and reds spread across the ground.

Golden Rowan of Menalowan waited for love's kiss,
But only the taste of acrid fruit would meet her parted lips.

How many berries will it take to find a perfect pair?
The bitter taste of failure may be more than you can bear."

"The rowan's fruit is not a berry, but a pome," corrected Lynette. "Nimue could never quite get her botanical terminology straight, even after Carl Linnaeus's explanations. We had to remind her of the difference between sepals and petals more than once."

"So, what do we have to do?" asked Elizabeth impatiently.

"Select the number of fruit by pulling out these knobs. That number of fruit will be dispensed when you spin the dial. Remember that you need a pair of the same color. Any fruit without a match has to be consumed." Lynette made a face.

"The bitter taste of failure?" asked Elizabeth. She despised certain tastes even more than unpleasant odors.

"Not just a bitter taste," said Lynette. "Legend says that eating too much Rowan fruit can kill you."

"Is that botanically accurate?" asked Elizabeth.

Lynette looked grave. "In a land of myths and legends, one shouldn't take unnecessary risks," she said. Before Elizabeth had a chance to study the fruit-dispensing machine, Lynette pulled out two knobs and turned the dial. Two pieces of fruit rolled into her palm. One was yellow and the other red.

Lynette sighed before popping them into her mouth. Her face scrunched up, and she swallowed quickly, gagging and unable to speak. Elizabeth tried to offer Lynette some water, but as she reached for the pitcher, Lynette's expression turned from distaste to pure fear. Elizabeth's hand dropped to her side.

"Your turn," Lynette croaked, barely managing to get the words out.

Elizabeth thought about the problem. *There are three different colors, and you need a pair.* She pulled out four knobs. Lynette's eyes grew wide. "The pigeon-hole principle," said Elizabeth confidently as she turned the dial with one hand and used the other to catch the fruit. One fruit was red, one was pink, and two were yellow.

Lynette patted Elizabeth on the back. *Was it in congratulations or in empathy for the task she had to do next?* Elizabeth placed the yellow pair on the bench and eyed the remaining fruit in her hand. She took a deep breath and closed her eyes, ready to take her medicine. When she opened her eyes, the fruit was gone, and Lynette was doubled over in pain.

"You swallowed the fruit for me," Elizabeth whispered in disbelief.

Lynette straightened up, held three fingers over her heart, and forced a weak smile. *As ridiculous as these ladies sometimes appear, they truly have hearts of gold,* thought Elizabeth. She put her arm around Lynette and wondered whether she would be equally loyal and selfless when the time came.

Elizabeth turned back to the pitcher of water. Sitting next to it on the bench lay a second note. Elizabeth read it aloud:

"*Golden Rowan of Menalowan shed tears that fell like rain,*
Only the flesh of hazel could make her whole again.
Golden Rowan of Menalowan sent thrushes to the skies
They brought back Fin MacCool, the boy with hazel eyes."

"This is a very different poem about the Golden Rowan of Menalowan than the one I remember," noted Elizabeth. "I thought Golden Rowan was a spinster."

Lynette, who had now regained her composure was standing behind her. "The famous one was about the first Rowan. This poem is about one of the others - perhaps even Nimue herself."

"And the boy with hazel eyes?" asked Elizabeth.

"Do you not know the legend of Fion Mac Cumhaill?" Lynette asked.

"The boy with hazel eyes?"

"Yes, although I suspect that he had hazel running through his veins, as well."

Things could always be worse, thought Fion Mac Cumhaill as he put a fish on the spit and moved it into the fire. Still, being a student of The Druid was not at all what he had expected.

Earlier that day at the market, Fion had joined a group of young apprentices who complained of the hard labor their masters had imposed. The boys had eagerly exchanged stories, each more impressive than the last.

One lad had chopped firewood for the entirety of a bitterly-cold night until his fingers bled. When the work was done, his master made spirits arise, black and howling, from the great fire.

Another boy was tasked with piling up enormous rocks for three days. His master, using powerful spells, had then raised a great serpent from the stones.

Fion had only listened in silence. When the apprentices had compared their rough, blistered hands, Fion placed his own soft hands in his pockets and sullenly walked away.

The pleasant odor of fish filled the room. Fion had come to study under the Druid master to learn the magical arts, but now he began wondering if he had made a mistake. The Master did not act at all like a real wizard. In the nine months Fion had worked for him, his master had never stirred a potion. He had recited only poems, never incantations, and not a single beast had been conjured. Instead, the Druid spent his days lecturing Fion on the great power of knowledge and training him in the art of poetry. The Master chopped his own wood, cleaned his own abode, and even cooked his own meals.

Very little was expected of Fion. Every even-numbered day, from sunrise to sunset, he was made to recall long epics under the Druid's watchful eye as they stood beside the clear water of the Pool of Fec. On the odd-numbered days, Fion was left to his thoughts while his master sat by the Pool, a hefty pile of books by his side. Ready for adventure, Fion's patience was wearing thin.

A burning smell suddenly hit the apprentice's nostrils. He jumped in panic and lunged at the spit with both hands. One hand successfully found its mark and turned the spit. The other hand landed on the skin of the fish as it was turning. Fion stuck his burned thumb in his mouth as he assessed the damage. The fish was scorched on one side. Fion shook his head, disgusted with himself. The Master had stressed the importance of cooking the fish properly, and now he had managed to foul up the simple task - the first real work assigned to him while in the Druid's service. But as Fion continued sucking on the finger that had made contact with the fish, something changed.

Fion lowered his eyes in embarrassment as he served the half-burned fish to his master. The Druid was silent. Fion finally summoned his courage. "My apologies," he said, looking up. The instant the Master's eyes met Fion's, he understood. The apprentice's eyes were no longer filled with the ignorance of youth; They were the eyes of a sage.

Instead of scolding the boy, the Druid rose from the table and gestured for Fion to take his seat. "The fish is yours to eat, my boy. The prophecy is fulfilled." Fion, a hungry youth, did not require further cajoling from his master.

As Fion devoured his meal, the Druid explained that it was no ordinary fish the boy was consuming, but the very Salmon of Knowledge that the Master had been seeking for seven long years. The salmon had grown in a secret pool where it fed exclusively upon hazelnuts. Prophecy claimed that the man who ate the Salmon imbued with the magical properties of hazel would become the wisest and most powerful being in the world. Fion accepted this honor with both relish and humility. He had always felt that he had a greater purpose. Now, his time had come.

The Druid, understanding that Fion's power was now greater than his own, bid him goodbye and good luck. As Fion walked away, he processed the new knowledge, borne of hazel, that had entered his body. He now realized that true magic lay not in tricks, but in knowledge. He understood that a verse held as much power as a hundred swords. And, most importantly, Fion accepted that the properties of hazel and the responsibilities of leadership must be received with a pure heart.

Fion paused at the Pool of Fec. A thrush alighted on a nearby branch and whistled a love song- an invitation. *How best to use my new knowledge?* wondered Fion as he listened to the harmonic intervals of the thrush's tune. Then he knew. To prove his wisdom, his authority, his mastery of the magical arts, Fion Mac Cumhaill, the most powerful being in the world, composed a poem.

"And now you understand the magical power of hazel," said Lynette.

"It sounds fantastic," admitted Elizabeth. "Knowledge, power-"

"It's extremely dangerous!" snapped Lynette. "Especially for one whose heart has even a drop of impurity."

Elizabeth was confused. "What was Fion's poem?" she asked.

"This is no time for poetry," scoffed Lynette. "We have a jewel to grow!" She lifted the pitcher of hazel water very carefully, uncovering a third piece of paper.

Elizabeth picked it up and read:

"Golden Rowan of Menalowan and hazel Fin MacCool
Together bore a magical seed – a time-traversing jewel.

Lay your berries in the ground and water them with care,
Will the seedlings of your love have any fruit to bear?"

Elizabeth led the way back to the garden. Lynette followed slowly behind, cradling the pitcher of hazel water as if it were a newborn babe.

Lynette, now caressing the pitcher, gestured to a spot in the dirt. Elizabeth dug a small hole in the soft ground. In it, she placed the fruit - two yellow spheres that glimmered in the setting sun. Elizabeth covered them with dirt. She reached to take the pitcher of hazel water from Lynette. Lynette, at first, would not loosen her grasp, but as her eyes met Elizabeth's she snapped out of her trance and nearly dropped the vessel into Elizabeth's chest. Elizabeth poured the water very carefully. The large drops hit the ground like small bombs. "Is that enough?" she asked.

Lynette shrugged. Soon their questions were answered. As they looked on in astonishment, a little plant pushed its way out of the soil. The stem rose skyward. Leaf after leaf uncurled. When the plant had reached the height of Elizabeth's knee, it stopped growing and... Pop! Pop! Two pea-pods appeared. There was a moment of stillness and a breath of anticipation. Lynette grabbed one of the pods, and Elizabeth the other. They opened them eagerly.

"Mine are all yellow," cried out Lynette in disappointment. "Not a jewel among them."

Elizabeth opened her pod to find four small yellow fruits which were identical in shape and hue to their parents.

"We need to try again," said Lynette looking over her shoulder. "But I'm not sure I can stomach any more of the 'bitter taste of failure' - I've already had four."

Elizabeth prepared herself for self-sacrifice. Suddenly an idea dawned on her.

"Did you say this place is called Mendel's garden?" she asked.

"That's what Nimue called it. Why?"

"Is it named after Gregor Mendel by any chance?"

"Now that you mention it, I do recall Nimue speaking of a certain Brother Gregor. She said he must have been a very hungry man as he dreamed of nothing but peas. This was in the early days when the jewels only let us eavesdrop on dreams. As her own research progressed, Nimue became more reclusive. She never mentioned Brother Gregor again. Do you know him?

Elizabeth smiled. "Now it is my turn to tell the story. Nimue was very lucky to get a glimpse of Gregor Mendel's genius, as he was a private person. Mendel was a man of faith - a priest, and later a friar, but he was also a student of mathematics and a serious scientist. The systematic diligence and precision Mendel applied to his experiments and the statistical acumen he used to analyze the results set the very foundation of modern genetics."

Lynette nodded, but her eyes were vacant. Elizabeth frowned. *Lynette was such a fabulous storyteller, and here I sound like an encyclopedia entry.* She tried again. "Gregor Mendel was a student of magic!" At this, Lynette perked up. "He had a delicate touch, patience that could uncover the magical pattern embedded in all living thing - even in the modest pea," Elizabeth continued.

"For the first act of Mandel's *Magic Show of Disappearing Traits*, he bred two different varieties of pea plants together - one with round peas and the others with wrinkled peas. In our world, plants are fertilized differently, and there was no hazel water available, so Mendel's magic took time to cultivate. But soon enough... Presto Change-o, the pea plants grew, and there was not a wrinkled pea among them! Where was the trait for wrinkled peas? Had it vanished forever?"

Lynette clapped lightly. Elizabeth, encouraged, continued.

"Mendel's magic act was not over yet. For the Second Act of the performance, Mendel took the hybrids - the new generation of round peas - and bred them with each other. Three-quarters of the plants in this third generation yielded more round peas, but... Hocus Pocus - the remaining fourth were - drumroll please... wrinkled peas!"

Lynette's eyebrows lifted.

"Mendel used these results to draw significant conclusions about how the magical elements that are responsible for visible traits - we call them genes - are passed on," explained Elizabeth. She had never used the word "magic" so many times in a scientific explanation, and it felt a little wrong.

"If the magic that turns a fruit into a jewel behaves like a recessive gene- umm, like the hidden magic for wrinkled peas - then maybe we can still reveal it." Elizabeth opened her hand and looked at the eight shiny yellow spheres. "Should we plant them?"

"Are you sure about this?" asked Lynette. "If we use up all the hazel water, there will be no more chances.

"I can't say I'm one hundred percent positive," Elizabeth admitted. "Even if I'm right about the rowan fruit's traits being carried in a similar manner to genes, there's still the chance that neither of the two original fruits had the 'magic gene' inside them to begin with. In that case, they can't pass the 'magic gene' to their offspring. It's a gamble, but I think we should try."

Lynette bent down and began digging little holes - four of them.

The two worked in ceremonial silence burying a pair of fruits in each hole. As the last drop of hazel water fell from the pitcher, the last ray of sunlight vanished.

Back in the laboratory, Lynette lit an oil lamp, and the eight pods were spread on the workbench. Pod after pod revealed more yellow fruit. "*Nature rarer uses Yellow than another Hue*,"[29] said Elizabeth mournfully, quoting Emily Dickinson. The irony of the words did little for her morale.

"Lady Emily would often say that," said Lynette, looking up.

"Emily Dickinson? Did she come here?" asked Elizabeth, keeping her eyes on the last pod.

---

[29] Emily Dickinson, "*Nature rarer uses Yellow than another Hue*"

"She visited Camelot in her dreams every night. Emily was a Lily Maiden herself and is even responsible for our organization's dress code," said Lynette, straightening out her white gown. "This garden was quite dear to Emily, and she would spend many days here among the buzzing bees, cataloging every plant. The squirrels adored her. She was the original Golden Rowan of Menalowa, you know. Her hair was dark, unlike the Golden Rowans that would follow, but she earned the name by virtue of her soul."

> "*She had the soul no circumstance*
> *Can hurry or defer.*
> *Golden Rowan, of Menalowan,*
> *How time stood still for her!*
>
> *Her playmates for their lovers grew,*
> *But that shy wanderer,*
> *Golden Rowan, of Menalowan,*
> *Knew love was not for her.*
>
> *Hers was the love of wilding things;*
> *To hear a squirrel chir*
> *In the golden rowan of Menalowan*
> *Was joy enough for her.*"[30]

"That sounds very much like Emily," said Elizabeth, thankful for the distraction. There was now only one pod left.

"In most legends, the last try is always the successful one," said Lynette hopefully.

Elizabeth cracked open the pod with shaky hands. She peered inside. The pod held four yellow fruits. "No

---

[30] Bliss Carman, "*Golden Rowan*", 1895

jewels." Elizabeth felt tears well up in her eyes. "I'm sorry," she whispered.

Lynette scanned the room for ideas. "Look at this!" she yelled out, pointing to a small tin cup. At first, it appeared empty, but as Elizabeth strained her eyes, she saw that it contained some liquid - perhaps only a few drops.

"Do you think it will be enough?" Elizabeth asked.

Lynette didn't answer, but Elizabeth knew what she must do.

She gulped and returned to the fruit-dispensing machine. *If only it held gumballs!* The four knobs were already pulled out, and everything was ready. Elizabeth suddenly realized that had she chosen to dispense six fruits at the beginning instead of four, she would have been guaranteed at least two pairs, with at most two unmatched fruits to swallow. But it was too late now.

> *"If you can make one heap of all your winnings*
> *And risk it on one turn of pitch-and-toss,*
> *And lose, and start again at your beginnings*
> *And never breathe a word about your loss;"*[31]

Elizabeth turned the dial.

Play Along: In this chapter's "play along" section, you'll explore the pigeonhole principle in more depth, learn about Gregor Mendel, the "father of modern genetics", read about the life and poetry of Emily Dickinson, and more. Flip to page 224 to join the fun.

---

[31] Rudyard Kipling, "*If—*", 1895

8. The Lake of Tears

"We're in luck," whispered Lady Elfinhart when they reached the lake. "Victoria will bloom tonight."

"Victoria is a lily?" gasped Lulu looking at the enormous round plant in the middle of the lake.

"Not at all! Victoria is a *water lily*, which is entirely different," corrected Lady Elfinhart. Leaning over, she whispered in Lulu's ear, "She's a dicot." Lulu looked puzzled.

Lady Olwyn touched her head to theirs.

> "*The lily sprouts a single leaf-*
> *a single truth that we hold dear,*
> *Victoria's seed holds two cotyledons —*
> *it's this duplicity that we fear.*"

Lulu tried to resist the urge to roll her eyes. She didn't want to insult the Ladies, but the flower symbolism was just going too far. *If I put such characters in my book would they be believable? Flower symbolism aside, can a mere mortal even be so honorable and pure?* she wondered. The eye-rolling urge was the first time that Lulu felt like her old self since coming to Camelot. It was a nice warm feeling- like turning from a shadow back to flesh and blood.

"But isn't Victoria the guardian of the magical jewels?" Lulu asked aloud.

"Which have mysteriously gone dull," added Lady Elfinhart, cocking her head to one side. "It appears that Victoria is preoccupied with her own affairs tonight. Watch!" She pointed to the gigantic water lily.

Lulu noticed that Victoria was proudly displaying a large white flower bud. As the bud opened, the smell of fresh pineapple (*or was it butterscotch?*) wafted through the air. Lulu took a deep breath and closed her eyes. When she opened them, a black beetle was crawling toward the center of the flower. The flower closed around it. *Was the beetle entrapped or embraced?*

"Is Victoria carnivorous?" asked Lulu, fascinated.

"She's offering a room for the night."

Although watching an over-sized plant devour a beetle held a certain appeal, Lulu was on a mission. "How do we find out what is going on with the gemstones?" she pressed.

"The Lady of the Lake will know. She lives on the Lake Isle of Innisfree."

"Nimue?"

"No, she was one of the Ladies of the Lake, but there were others."

"Arthurian legends are very confusing."

"That is the nature of the quilt of Camelot, stitched by many hands over many centuries."

> "*When life lives as legend under the guise*
> *Of reality viewed through dream-veiled eyes...*"

Lady Olwyn's voice faded. Lulu waited for the rest of the song, but there was no more.

Lady Elfinhart waited until she had Lulu's full attention. "Listen carefully," she said, her tone severe.

89

"The current Lady of the Lake is no other than the Fairy Queen herself. She announced the appointed shortly after Nimue's disappearance and now refers to herself as Q".

"Did you say 'Q'?" Lulu asked. *Where have I heard that name before?* A faint dream-like memory suddenly appeared like a seedling that had pushed its way up through the dark soil. *Petrus!* Lulu reached in her pocket and felt the neatly-folded scrap of paper (which she knew depicted Petrus's invention). There was no need to share it with the Ladies.

Victoria neared the shore. "She will take you to the Isle of Inisfree," said Lady Elfinhart, pointing to the giant water lily. "But be careful," she warned. "Remember that when you deal with Fairies, there is always a price to pay."

Lulu nodded in understanding. She put a timid foot on Victoria, unsure whether the plant would support her weight. It did. Soon she was drifting across the lake, and the Ladies disappeared from sight. It was then that Lulu heard singing. Hundreds of tiny haunting voices surrounded her. They strung the words of William Butler Yeats into an eerily beautiful melody:

> "*Come away, O human child!*
> *To the waters and the wild*
> *With a faery, hand in hand,*
> *For the world's more full of weeping*
>   *than you can understand.*" [32]

A small delicate hand reached from the water, beckoning. Lulu knelt and looked over the side of the water lily, trying to get a glimpse of the hand's owner.

---

[32] William Butler Yeats, "The Stolen Child", 1889

She had never seen a Fairy before. As Lulu shifted her weight, another hand popped out of the water and grabbed her by the wrist. A third wrapped around her ankle. Lulu, terrified, pulled away. The delicate hands lost their grip, but new hands quickly took their place, gently stroking her arms and legs. Lulu moved to Victoria's center and rolled her body into a ball. The singing grew louder:

> "*Come away, O human child!*
> *To the waters and the wild*
> *With a faery, hand in hand,*
> *For the world's more full of weeping*
>   *than you can understand.*"

There was a desperation in the voices. They were pleading, promising, persuading. Lulu covered her ears, but the song persisted. As Victoria moved closer to the shore, Lulu leaped toward the safety of solid land. She turned back to see that all the Fairy hands had vanished. Lulu bowed her head to Victoria in gratitude. Whether Victoria was merely a plant or a sentient being, it seemed like the right thing to do.

The Isle of Innisfree was an even smaller eyot than Lulu had imagined it would be. Long and narrow, the island resembled a boardwalk. A single dilapidated wooden cabin lay at its far end. The remains of a neglected garden and abandoned beehives sat at its margins looking mournful in the moonlight.

Lulu had a tough time imagining a Fairy Queen choosing to live in a rustic cabin. Following what must have once been a beautiful white stone walkway (but was now overrun with weeds), Lulu timidly approached the dwelling. "Enter," came a voice from within.

The Fairy Queen was the most beautiful woman Lulu had ever seen. She shone like a moonbeam and a streak of light trailed every one of her movements. The gown she wore came up to her neck and was covered with tiny iridescent hummingbird feathers. Somewhere in the back of Lulu's mind, Lady Elfinhart's warning rang out. With excruciating difficulty, Lulu lowered her gaze from the stunning creature. Her eyes fixed on the floor.

It was there that Lulu saw N. Wake. Their eyes quickly met as the green mouse disappeared into a sea of tentacles. "I wished for feet like water lilies, and this is what I got!" growled the Fairy Queen. "I would fry that fish if it weren't for - ." The tentacles disappeared into folds of fabric.

"A bed!" exclaimed Lulu. She had lifted her gaze off the floor and was staring at the Fairy Queen in her full splendor. The Queen was glowing as brightly as ever, but Lulu was amazed at how anger had twisted her beautiful features into something repulsive. What Lulu had initially assumed was the Queen's throne was a bed. She instinctively looked at the headboard and was met with the radiance of a glowing jewel. "How?" Lulu managed to mouth.

"Have a seat, my dear," spoke the Queen. Her body glided aside, and she pointed to the space on the bed next to her. The Fairy Queen's face had softened and regained its beauty. "All your questions will be answered," she sang in a voice as smooth as honey.

Lulu remained standing. The Queen reached out and began stroking her hair. "Dear child," she murmured. "Dear, dear human child."

Lulu tried to pull away. "What is happening to the gems in the lake?" she asked quickly. "And why is the one on your bed still glowing?"

The Queen answered Lulu's interrogation with another question. "What does a woman most desire?" she asked.

Lulu was not sure how this related to the jewels. "A woman most desires… an answer to her question," she ventured.

"What does a woman most desire?" the Queen repeated, raising her voice.

Lulu studied the Fairy Queen's face. It was the epitome of perfection.

"Beauty?" Lulu tried, hoping to appeal to the Queen's vanity. The Queen gave no indication that she had heard Lulu's answer at all.

"What does a woman most desire? Such was the riddle posed to King Arthur by Sir Gromer. The knight, vengeful over a land dispute, agreed to spare Arthur's life in exchange for an oath. Arthur swore to return in a year bearing the correct answer to the riddle. Should he fail, Arthur's life would belong to Sir Gromer."

Lulu was intrigued by the story. As she listened, she no longer pulled away from the touch of the Queen's hands. The Fairy Queen, seeing her opportunity, drew Lulu to her side as she continued the tale.

"The sun rose and set. The moons waxed and waned. The deadline was drawing near, and the answer remained elusive. Soon there was but a single day remaining before Arthur would have to face his fate."

"Why not run away?" asked Lulu. "Why not gather an army and defeat the riddle-poser?"

The Fairy Queen smiled, and her long fingers continued to work their way through Lulu's hair. "Those would be the actions of a wise man. But King Arthur? Arthur kept his promises in the name of honor. Bah!"

The Queen's face distorted again and she no longer looked beautiful.

"Arthur, in his foolishness, had resigned to his destiny. As he rode to meet Sir Gromer with one of his knights, Sir Gawain, at his side, the mood was somber. No sooner had the two stopped in a clearing when a woman appeared from the woods and walked toward them. As she approached, Arthur and Gawain could see that she was a loathly lady, indeed. Her clotted hair hung down over a bright red face with protruding teeth. The woman hobbled toward them, a great hump upon her back.

> *"She was as ungoodly a creature*
> *As ever man saw, without measure."*[33]

The Fairy Queen described the loathly lady with such passion and such disdain that Lulu wondered if this was someone she had known personally. *Or, perhaps, an image of a former self?*

"Arthur and Gawain recoiled at the creature's ugliness and made ready to mount their steeds and ride away. 'I advise you to speak with me, Arthur,' cackled the hag, 'for your life is in my hands.'"

"Arthur froze. 'What do you want from me?' he asked. The loathsome lady revealed that she was Sir Gromer's sister. 'The answer to the riddle could be yours,' she croaked."

"Arthur hesitated. He wondered for a moment if the woman was not a witch and whether her intentions would prove as ugly as her visage. It was as though the woman was reading his thoughts."

---

[33] ANON, *"The Weddyng of Syr Gawen and Dame Ragnell for Helpyng of Kyng Arthoure"*, 15th century

"Forsothe," said the Lady, "I am no qued [34].
Thou must grant me a knight to wed:
For it must be so, or thou art but dead;
Choose now, for thou may soon lose thine head." [35]

She was looking straight at Sir. Gawain."

"Did Sir. Gawain agree to marry the witch?" asked Lulu eagerly.

The Queen nodded. "In the name of honor," she hissed and rolled her eyes back into her head. The whites of her eyeballs showing, the Queen was not just hideous – she was utterly revolting. Lulu made a mental note never to roll her eyes again.

"And the answer to the riddle?" Lulu asked.

"What does a woman most desire?" The Fairy Queen's face softened and in an instant regained its former beauty.

"What does a woman most desire?" she repeated, shaking her great feathery shoulders. "You desire to know what is happening to the jewels."

"But the riddle-" Lulu began.

The Queen raised her voice. "Do you not desire an answer to your questions over Sir Gomer's?"

Lulu, feeling miserable, nodded.

The Fairy Queen sighed. "The world's so full of weeping," she said mournfully. Her feathered gown was now brushing against Lulu's cheeks. "Dozmary Pool – 'The Lake of Tears' - is growing saltier by the day. When the jewels lay at the bottom of the lake, Victoria could energize them and protect them. Feet of water lilies, indeed!" The Queen's face distorted for a moment.

---

[34] In Middle English, a qued means an evil person.
[35] "The Weddyng of Syr Gawen and Dame Ragnell for Helpyng of Kyng Arthoure"

"Now they're in a state of neutral buoyancy - suspended in the water like stranded shipsssss," she hissed, pulling her serpent-like tentacles from beneath the covers and waving them in the air. "But soon-" the Queen howled. "Soon they will float to the surface to be collected and placed into my tender care." The hand on Lulu's hair was stroking faster now.

Lulu suddenly remembered Lady Elfinhart's warning and jerked way from the Queen. She caught herself off guard and landed on the hard floor. The shock of the fall quickly gave way to the surprise of seeing N. Wake. The green mouse winked at Lulu and put a paw over her mouth. It took Lulu but a moment to understand. It was time to take decisive action. The Ladies has entrusted her with this important task. Lulu rose to her feet and smiled politely at the Fairy Queen.

"You must know how to take care of jewels better than Victoria," Lulu said sweetly. The flattery worked, as the Queen beamed and a halo formed around her face so bright that the glare momentarily blinded Lulu.

"My jewel is guarded by the Fish of Desire," responded the Queen. Her human hands pointed to the headboard of the bed while her tentacles vanished beneath the covers once again. Lulu noticed a small glass bowl filled with tea-colored water. At its bottom lay a jewel. A small fish was circling protectively around the treasure.

"The fish swims in hazel water. That's the source of its magic," said the Queen, tapping on the bowl with a fingernail. The fish swam faster.

Burning fire,
Hazel, Rowan,
The fish of desire,

The price of knowing.

"What ever happened to the other Ladies of the Lake?" Lulu asked quietly. "What happened to Nimue?" She immediately realized her mistake. The tentacles re-emerged and waved threateningly in the air before seizing Lulu by the ankles. Before Lulu knew what was happening, she was thrown onto the bed.

"We'll need to prepare for the rest of the jewels," said the Fairy Queen nonchalantly as if nothing had happened. "More hazel water is required. And a larger container. Are you with me?" Her human hands lifted the fishbowl to her lips.

Lulu, using great restraint, pulled the corners of her mouth up in an expression that she thought might resemble a smile. "Of course, Your Majesty," she said through clenched teeth.

The tentacles loosened their grip, and the Queen returned the bowl to its place on the headboard. A drop of hazel-water glistened on her lips. "Then off to the fountain we go," the Queen announced. She released Lulu and licked the drop with a long dark tongue. The Fairy Queen positioned herself in front of the headboard and, using both of her hands and at least a dozen tentacles began pushing buttons at a feverish pace.

Every time Lulu leaned over to have a closer look, a tentacle would slap her away. She wasn't sure she would have observed much, anyway, as fingers and tentacles were moving too rapidly. In a moment, the clicking of the keys stopped, and flaccid limbs dropped to the bed. The Queen's limp body followed, landing on top of Lulu.

"Push her off!" squeaked a voice. Lulu put her arms around the sleeping Queen's middle and heaved.

The Fairy Queen's body rolled to the edge of the bed. Lulu repositioned herself for the final push. *Had the Queen's body grown heavier?* Thrusting now with her entire body weight, Lulu felt herself only sink into the softness of the bed. Her arms were so tired, *so very tired...*

Play Along: Learn more about the Victoria plant, decipher Middle English in "*The Weddynge of Syr Gawen and Dame Ragnell*", and explore the principle of buoyancy with a floating egg experiment. Join the fun on page 232.

9. The Pool of Joy

When your soul holds joy like a fountain,
A single drop can move a mountain.

The words of the dream echoed in Lulu's head.
The Fairy Queen was gone. *Did I push her off the bed?* The
Fish, too, had disappeared.

The bed was situated next to a new body of water.
Unlike the stagnant Lake of Tears, however, this pool was
alive- flowing, dancing, giggling. Bathed in moonlight,
the water bubbled in reckless abandon- as if boiling
without steam. Looking upon the "Pool of Joy" (as she
had already named it), Lulu felt a sense of new hope and
delight.

The sparkling pool was fed at one end by a spring
that trickled steadily from a jagged cliff-face. At its
center, an enormous rock protruded from the water. On
this stone, the likeness of a knight and lion stood on
guard. "Sir Yvain and his lion!" Lulu said aloud, recalling
the Arthurian legend. The Lion bowed his head, and
Lulu jumped back. (It was only later that she would see
the mechanism that made this movement possible - a
simple water wheel hidden in the lion's metal tresses.)

Lulu leaped off the bed full of excitement. She
stepped toward the pool, ready to explore, but a strange

new feeling held her back. "I must be prudent," Lulu advised herself. For a moment, she wondered if the voice was her own and not her sister's. *Why hadn't Elizabeth and Lynette met them at the lake?*

Lulu shrugged off the worry with a lonely sigh and set to the first order of business - figuring out how the bed worked. *If I need to make a quick escape, I better know how to operate the bed,* she reasoned.

Lulu inspected the entire bed, as well as the striped bedsheet, but couldn't locate any written instructions. The left-hand side of the headboard held a large toggle switch. One side was labeled '*HERE*' in ornate silver letters, and the other side was marked '*THERE*' in gold. The switch was currently pointing to the latter. Lulu would have typically made a joke about being 'there' and not 'here', but the thought of her sister still lingered. She continued her task with an air of solemn determination.

A red number '3' was painted next to the switch. Embedded in the wood to its right, an emerald lit up the bed with its green brilliance. Following the '3', there was a succession of small round slots. The first two slots were filled. Each one contained a single marble held in place by a pin. Lulu carefully pulled out the first pin and turned the marble in her hand. It had the number '1' written on it. Using equal care, she examined the second marble. This one was a '4'.

The rest of the marbles and pins had fallen out during the voyage and were sitting in a long container presumably created for such a purpose. Lulu estimated that there were at least fifty marbles. "So many combinations," she sighed and leaned against the collection shelf. The shelf broke beneath her weight, sending marbles and pins scattering all over the bed.

As Lulu began cleaning up her mess, she suddenly stopped and smiled. The pins had spread on the striped bedsheet. Some were intersecting a stripe, while others lay between them.

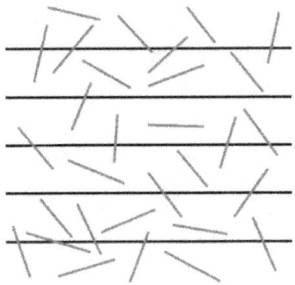

The image of the pins that had fallen on the striped bedspread awoke an old memory – an experiment she'd done with her mother. They'd dropped a needle onto a piece of lined paper over and over. When the total number of needle drops was multiplied by two and then divided by the number of needles that had hit a line, they got a special number- Pi. "And that's Magic!"[36] Mom had whispered in her ear. *Mom obviously knew nothing of real magic,* thought Lulu. *But she does know a thing or two about math.* And for this, Lulu was grateful.

"Easy as Pi!" she yelled out. Lulu's voice bounced off the walls of the cliff and echoed across the water.

Soon the bed was all set up with the numbered marbles displaying the first fifty digits of Pi. All Lulu would have to do, should she have to leave in a hurry, would be to toggle the switch to *'HERE'*. She felt proud of herself for tackling responsibility first and adventure second. *I must be growing up,* she thought to herself. This

---

[36] It's really called Buffon's needle problem.

realization filled her insides with an odd mixture of excitement and terror. "The time for exploration is upon us," said Lulu, capturing Lady Elfinhart's tone perfectly.

Lulu ran to the pool and stopped in front of the tallest pine. Fastened to one of the tree's high limbs was a long metal chain connected to a large iron basin. Beside it, sat a smaller, free-standing basin. Lulu examined the two vessels.

The number 5 was painted inside the larger basin and the number 3 inside the smaller one. The chain was long enough to allow the larger basin to be filled in the spring.

The intricate designs on the outside of the containers caught Lulu's eye. She lifted up the larger bowl to have a closer look. A painted chain of lions circled its base, each with its tail inside the mouth of a dragon. Beneath the lions and dragons, Lulu could make out some writing. She turned the vessel and read:

> *With a pure heart, this basin you may borrow.*
> *Exactly four parts will dilute all the sorrow.*

Lulu noted the irony in the fact that the basin that could be 'borrowed' was chained to a tree.

Lulu examined the smaller container. It was covered with an irregular pattern of dots and dashes. She copied the pattern in her notebook in case it would prove important.

-- * * *, *, *-- --, * --, *--*, *.

Lulu suspected that the message was written in Morse code, but didn't have a key to use for decoding. *If I put this in my book*, Lulu thought, *my smart, resourceful readers will decrypt it.*

Lulu lifted the first bowl, this time using the chain, and lowered it into the pool. When the basin was full of water, it was quite heavy, and Lulu wished it was attached to a pulley. "I have five units of water now, but I need exactly four," said Lulu aloud. She pulled her hands into the sleeves of her shirt to use as mittens and carefully poured the water into the smaller basin until it was full. After seeing what the hazel-water had done to the Fairy Queen, Lulu didn't want to take any chances.

"Now I have two units in the larger basin," she declared. "Filling, pouring, and a touch of discrete mathematics and I'll have the job done in no time!" Using her gloved hands, she picked up the smaller container and emptied it back into the stream. Something shiny in the water caught her eye. For a moment, the moon's distorted reflection had two red eyes. Lulu leaned in for a closer look. Two polished rubies lay at the bottom of the stream. All at once, the rubies were surrounded by sparkles of light.

"Well done, my dear!" The Queen's voice startled Lulu. She lost her balance and had to flap her arms wildly to keep from falling into the water. Lulu spun around on her heels and found herself face to face with iridescent feathers. The Fairy Queen was standing at her full height, beautiful and shining more radiantly than the moon itself. She was holding the fish in its bowl. The Queen pointed to the rubies. "Retrieve them!" she commanded, her voice rising sharply.

Lulu stood motionless. She knew her shirtsleeve would be immediately soaked if she reached her arm into the magical waters. *Maybe I can use the basin with the chain.* Lulu placed the smaller container on the ground and reached for the larger one, thinking she'd transfer the water, first. Before she had a chance to do so, the Fairy

Queen shoved the fishbowl into Lulu's chest. "Useless child!" she huffed and headed into the water.

The water boiled around the Fairy Queen's tentacles. Without hesitation, the Queen's arm disappeared beneath the water and re-emerged holding two perfectly round rubies. The Queen marched from the pool and dropped the jewels into the fishbowl. She licked her long fingers with her even longer black tongue. As the rubies plopped into the fishbowl, a drop of water fell onto Lulu's sleeve. Where the water had touched her shirt, the fabric turned to gold.

"The water for our new jewels can wait," boomed the Queen, kicking over the smaller basin. "We have wishes to make!" She winked at the lion statue, snatched the fishbowl from Lulu, and spun toward the pine tree. The lion bowed its head, but only Lulu was watching now.

When Lulu turned back to the tree, the Fairy Queen had vanished. Lulu walked carefully around the base of the pine and allowed her eyes to adjust. Suddenly, she felt a tentacle grab her forearm and pull her into an opening in the tree. Lulu rubbed her arm where the tentacle had pulled and groped blindly through the darkness. Something was exciting and familiar about stepping into a tree. *Maybe I'll end up in Wonderland,* Lulu hoped.

Lulu's shin bumped into something hard, which further inspection revealed to be a staircase. The steps were narrow and steep, so Lulu used both hands and feet to climb. The staircase wound around and around inside the tree. Lulu looked up and thought she saw a hint of light. *Was it the moon?* The light grew brighter as she ascended. The stairs came to an abrupt halt. Lulu had to crawl on her belly onto the last step. She saw that this level led straight out of the tree onto a balcony.

The Fairy Queen was already on the landing, staring out across the moonlit pool. Without looking back, the Queen motioned for Lulu to join her. Long thin fingers and long thick tentacles waved in the air. Lulu continued crawling until she was out of the tree. Standing up, she finally had a sense of how high she had climbed. The wind felt cold on her cheeks. Lulu shivered.

"Help me with the crank," the Queen said. It was an order, not a request. Lulu saw the device that was attached to the balcony. The Fairy Queen was pulling on a large crank with one hand while the other held the fishbowl close to her feathered body. Lulu grabbed onto the end of the crank behind the Queen and hung on with the entire weight of her body. The crank groaned and began moving, reluctantly at first, and then spinning quickly on its own. There was a narrow railing around the balcony. Lulu peered over this railing to see what was happening. Two wooden slides were moving away from the tree toward the pool. The crank only stopped spinning when the base of both slides had reached the rock.

"When Sir Yvain first came to this fountain, he poured the magical waters over that rock causing a tempest to rage," said the Fairy Queen, covering Lulu with the warmth of her feathers. "He had not yet met the Lion, and when an enraged knight appeared, he had to fight him on his own."

The Queen began stroking Lulu's hair with her long fingers. "Have you ever dreamed of becoming Queen, Lulu?"

"Yes, often," Lulu admitted.

"As Queen, all these lands could be yours." The Fairy Queen swung her arm dramatically, and her tentacles swayed like cobras.

Lulu surveyed the landscape where dark forests spread out far across the horizon. "Isn't there a King now? After Arthur, I mean," said Lulu.

"King Hallden," sneered the Queen. "Whoever heard of ruling a kingdom through righteousness?" She spit into the wind. "What we need is a warrior, not a diplomat," the Fairy Queen scoffed. "A leader of Camelot must rule with fear and prove his worth with bloodshed."

"Young King Hallden has left Camelot weak and vulnerable," the Queen continued. "Even as we speak, armies are uniting against him. It will not be long before Camelot falls and power is restored to the one who-". The Fairy Queen's voice faded into the wind, and she did not finish her sentence. Lulu stood shivering, wondering why she'd ever followed the Queen up here.

"I have been cursed," the Fairy Queen suddenly moaned. "For half the day, I must turn into a horrible beast - a loathly lady, if you will." She turned to face Lulu, and her eyes were glowing. "And now you have a decision to make. Would you prefer that I make this transformation by night... or by day?"

Lulu looked away, and her eyes caught a thin strand of the sky in the eastern horizon that was a shade lighter than the rest of the heavens. It held the promise of morning. Lulu took her time to think as tentacles began lashing restlessly at her back.

"The choice is not mine," she said at last. "It is yours. Every woman has the right to make her own decisions."

"Well said," laughed the Fairy Queen. "What a woman most desires… is sovereignty."

"The answer to Sir Gromer's riddle!"

"You will make a wonderful Queen, Lady Lovelace."

Lulu wasn't sure how to reply. She frowned and put her hands in her pockets. Suddenly remembering, she pulled out the paper depicting Petrus's invention and held it out to the Fairy Queen. "From a friend-" Lulu began.

The Queen snatched the paper from her hand, unfolded it, and held it up to the moonlight. In a moment, the sky boomed with the Queen's rolling laughter. "Parabolas are completely out of fashion," she shrieked. "It is the cycloids who reign supreme." The Fairy Queen crumpled the paper and flung it into the wind. *Poor Petrus*, Lulu thought

A ray of light appeared in the East. "The time for wishing in upon us," cried the Fairy Queen. She reached into the fishbowl and pulled out the emerald and one of the rubies. The Queen pressed the emerald into Lulu's hand which, fortunately, was still protected in her shirtsleeve. The sleeve turned to gold where the drops of water had touched it. "These two slides will guide the gems to the Wishing Stone. Only the wish that arrives there first will be granted," the Queen explained.

Lulu studied the slides. Both of their paths began at the balcony railing with a wooden starting gate that would hold the jewels in place. The slides both ended at the large stone in the stream. One slide's path, however, pointed directly to its destination, while the other curved, starting with a steep drop and then scooping back up toward the rock. "Which path do you choose?" asked the Queen, her voice turning to velvet.

Lulu placed her emerald on the slide that led directly to the Wishing Stone. *A straight line is the shortest path, and the shortest path will be the fastest,* she thought. The Queen placed her ruby at the start of the other track. She pushed the fishbowl at Lulu. Lifting her arms to the moon, the heavens echoed with her voice:

> *"Desire burning like a thirst,*
> *Fulfill the wish that finds you first!"*

The Fairy Queen lifted the gate, and the gems began their descent. Lulu looked over the railing. Two glowing lights - one green and the other red - sped toward the rock. As expected, the Queen's gem had the lead in the beginning. Lulu held her breath and wished as hard as she could. The emerald on its straight path was gaining ground but in the end... it was not enough. Lulu watched in horror as the red light – the Queen's ruby - reached the Wishing Stone first.

The Lion rose onto its hind legs, and Sir Yvain lifted his sword toward the sky. Immediately, the moon disappeared, plunging the world into darkness. The entire forest boomed. Lulu fell to the ground and, still holding the fishbowl, scooted back toward the tree. As lightning flashed in the sky, she could see the silhouette of the Fairy Queen, her head swung back, and her arms still outstretched like wings. Lulu thought she spotted a small streak of green perched on one of the Queen's tentacles before the darkness once again closed in. When the next bolt of lightning lit the sky, the Fairy Queen was gone.

Play Along: In this chapter's "Play Along" section, you will find pi with Buffon's needle, test the Brachistochrone curve to see if a cycloid is faster than a direct path, solve "fill and empty" puzzles, read the Arthurian Legend of Sir Yvain and the Lion, and more. Join the fun on page 238.

10. Why the Aspen Quakes

"The Moon was but a Chin of Gold
A Night or two ago--
And now she turns Her perfect Face
Upon the World below--"[37] -Emily Dickinson

The moon was beaming at Elizabeth, setting the two amethysts in her hands aglow. The taste of the bitter fruit lingered on her tongue - a reminder that all victories come at a price.

"Now that we know how to grow gemstones, we can make Nimue's vision come true," said Lynette excitedly. "Imagine expanding our fleet of beds. Oh, the things we'll learn! Oh, the places we'll go!"

"Did Dr. Seuss visit Camelot, too?" Elizabeth asked in earnest.

Lynette's baffled expression answered the question. Elizabeth turned away, her face growing red.

"The question of the jewels' energy remains, however," continued Lynette. "Your jewels are glowing with the energy of youth, but it is only a matter of time before they'll fade like the others."

---

[37] Emily Dickinson, "*The Moon was but a Chin of Gold*"

Elizabeth took another look at her amethysts. "We have to find Victoria," she said decisively. "Lulu and the Ladies are waiting for us."

"All in due time," responded Lynette. "We have one more stop to make. This one is significant. It will change everything."

Elizabeth wanted to protest - to tell Lynette that she'd find her sister and the others on her own- but she was a stranger in this land. Elizabeth had no choice but to follow.

From the hilltop garden, Elizabeth could see a storm raging in the distance. Lightning danced to the clap of thunder. "Is it coming our way?" Elizabeth asked, pointing.

Lynette shook her head. "It's a magical storm - see the green outline around the lightning? They're usually localized."

The walk down the hill appeared more treacherous by night. Lynette held the oil lamp above their heads as roots grabbed at their legs.

The aspens shuddered in the breeze, whispering. As Elizabeth and Lynette approached the bottom of the hill, they could make out human voices. Lynette quickly blew out the light. On hands and knees, the two crawled along the trail, staying close to the bushes. They hid behind the linden tree as the voices grew louder and more distinct. Elizabeth pressed her back up against the trunk of the tree, but Lynette peered out.

Three men were squatting beside a fire they had built on Hengoen's back. Their horses were tied nearby, and Lynette's horse was nowhere to be seen.

"Nothing like fried pigeon to fill an aching belly," said one of the men. He belched loudly.

"There's sure to be more at the camp," said another. "They say hundreds of troops are amassing in the valley by the day." He dug up a stone and flung it. The rock hit the ground with a clank.

"The Black Pig sent word to all the neighboring kings," spoke the third. "They're arriving in units of a hundred men, trained and organized if you'd believe it. Two nights ago, I spoke to a man who saw it with his own eyes." He removed a flask and began gulping its contents.

"He said the units lined up three abreast and the valley looked like a dark serpent," the third man continued excitedly. "Two units were left behind in this formation and didn't like it none too well. Their commanders rode up to the Black Pig and words were exchanged. The man with whom I spoke said that the Black Pig would have done away with those commanders if it weren't for one of his advisors - a magician, I believe- who reminded him of the higher purpose ahead." The speaker coughed.

"With a new order, the units moved into lines of seven, and the valley transformed into a dark wolf. This time, three units remained. Whispers passed from ear to ear, and the Black Pig seethed with rage. This time, it was the Magician who stood before the troops, his staff raised, shouting incantations. By the forces of a supernatural power, the units moved like a great sea pouring from a flask. When the Magician lowered his staff, the units were lined up ten across, and none were left out."

"Camelot hasn't a chance," laughed the second man throwing another rock.

Lynette did not move, but Elizabeth could see her fists clench.

The first man looked apprehensive. He chiseled at the ground with the edge of his knife. "Are you sure we'll blend in with these troops?" he asked.

"There are the units, yes - too orderly for our sort," chuckled the third. "But there are always men in the peripherals- savage men...and beasts who answer to neither commander nor Wizard but only to their thirst for blood."

"And don't forget the giants. There will also be giants," added the second.

The third man gave him a shove. "We'll fit in among these men, eat our fill, take what's left about, and be off by nightfall tomorrow."

Lynette and Elizabeth crawled on their bellies down the hill. When they were safely out of earshot, Lynette let out a growl. "We have to tell King Hallden," she said. "And I'm afraid the messenger pigeon was turned into dinner."

"Is there another way?"

"Populus Post. It's not as direct as a messenger pigeon, but it should work."

"And where do we find a mailbox?" Elizabeth asked.

"The populus tremula is both mailbox and messenger."

"An aspen tree!" Elizabeth exclaimed. She picked up a yellow leaf from the ground and looked up, smiling. Not far away from them, stood an aspen grove.

"Most consider the mighty oak the giant of the forest, but none compare to the aspen," Lynette explained as they moved toward the grove. "What appears an aspen grove, is a single being, the shoots all connected beneath the ground. Thousands of years old, the aspen is the forest primeval itself. It reminds us that strong roots are everything.

The wind blew, and the leaves of the aspen trembled as Lynette and Elizabeth approached.

"Do you know why the aspen quakes at the slightest breeze?" asked Lynette.

"Actually, I do," answered Elizabeth proudly. "Its leaves are attached to the branches by flat petioles."

"It trembles," corrected Lynette, "because it is full of secrets. It shakes in fear of Merlin."

"Merlin the magician? I know the legends!" said Elizabeth.

"You know nothing!" responded Lynette, raising her voice more than was prudent. "Nothing but lies! Now climb to the top of that rock and watch for Merlin and the Black Pig's spies while I send warning to King Hallden by Populus Post. She bent down toward the base of the tree and placed a heart-shaped leaf of a basal shoot in one palm while reaching for something in her sack with the other. Sensing Elizabeth's eyes still upon her, Lynette turned around. "Go! Now!" she commanded in a loud whisper. "If Merlin intercepts this message, all will be lost!"

Elizabeth was filled with curiosity about the mechanics of Populus Post but did not dare disobey Lynette. She turned to the rock where Lynette had pointed and began climbing.

"It is done!" announced Lynette when Elizabeth was half-way up.

Elizabeth turned around, shrugged, and once again joined her companion.

"I'll fill you in on Merlin as we walk," said Lynette. "The legends you've heard, no doubt, paint a picture of Merlin in a favorable light that could not be

further from his true nature. Many in Camelot believed
his false facade, as well. Only Nimue and the Lilly
Maidens knew the evil of his heart."

"Merlin colluded with the worst types of fairies,"
continued Lynette. "Perhaps even the Devil himself," she
added in a whisper. "He appeared a loyal adviser to
Arthur, but secretly used his spells and incantations for
his own nefarious intentions."

Elizabeth still couldn't shake the image of Merlin
as a kindly cartoon wizard. "How do you know that's
true?" she asked, stomping her feet into the ground as she
walked.

"I don't - not for sure. But Nimue - she knew."

*"My mouth shall speak of wisdom; and the meditation of
my heart shall be of understanding."*[38] -Psalm 49:3

Nimue awoke, aware only of the pounding in her
chest. *How the dreams of The Monk differed from those of The
Poet!* Nimue had been communicating with Gregor
Mendel for almost a year now, but only recently had she
begun to understand the true nature of his hunger. To
Mendel, the peas did not represent a hunger of the
stomach at all, but a longing of the soul - a desire for
knowledge. This type of hunger must have been
contagious, for Nimue, as of late, could think of nothing
but her work.

Nimue stepped onto the floor, retrieved a box
from beneath the bed, and eagerly opened its lid. The
smell of freshly-turned earth immediately hit her nose as
she verified that her gardening tools were in order.

---

[38] Psalm 49:3, "*King James Bible*"

Nimue placed the box on an adjacent bed and quickly entered the code, one inspired by Mendel himself. The rowan fruit had left Nimue weak and rendered her left arm limp and useless, but her work - her discoveries - had restored her spirit.

"From one dream to another," Nimue mumbled drowsily.

But this time, there were no dreams. Nimue opened her eyes to the terrifying sight of the Fairy Queen standing over her. In her hand, the Queen held a flask of hazel water. Nimue startled, and her box fell to the floor. She feigned a low bow, then scooped up her tools. The Fairy Queen stomped a large foot on her finger.

"Mother," said Nimue, rising (this is how a Fairy Queen is to be addressed by another Fairy). "To what do I owe the pleasure of this visit?" She did not look the Queen in the eyes but leaned over to see whether anything was out of place in her private laboratory. The Fairy Queen stepped closer, and Nimue gagged on a mouthful of feathers.

"I'd like to introduce you to a dear friend," spoke the Queen. She stepped aside to reveal a robed man who was examining the rowan-fruit dispenser. "Merlin, if you will."

The man turned around, and Nimue immediately recognized the Wizard. Merlin lifted Nimue's hand to his face and kissed it. It was the arm which was already numb, and Nimue did not feel a thing.

Nimue lowered her eyes. She knew what was expected of her by the look on the Queen's face. After the Fairy Queen had sent her to the Poet (*William was his name*), Nimue vowed never again to allow herself to be turned into the Fish of Desire.

"I'm no longer the Lady of the Lake," whispered Nimue. "What do you want from me?".

The Queen stomped on her delicate toes.

"I shall step outside now," she said. "I'm sure you and Merlin have much to discuss."

The Queen tromped out the door, and Nimue was alone with the Wizard.

"You and I," began Merlin, "are both students of the magical arts. He moved toward Nimue.

I'm not a wizard, but a scientist- Nimue thought, *-an adventurer, a seeker of truth.* She kept her lips tightly pursed.

"The Fairy Queen has told me of your wonderful discoveries," Merlin continued. He opened his hand to reveal two black gems. "I have been working on some experiments of my own."

Nimue clenched her fists. "I'm sorry, sir, but I'm afraid the Queen has exaggerated. My abilities are naught compared to a powerful Wizard's - mere amusements of a Lady."

Merlin raised a finger to his brow. "Shall we walk?" he asked. "I have many treasures to show you."

Nimue took a step back. "I'm sorry-" she began.

"You will surely be interested in these relics. There's a bed, a book, and a horn."

Curiosity overtook Nimue. "A book?" she asked.

"The Book of Giants," Merlin answered.

"And the bed?"

"A bed large enough for a giant himself."

Nimue swallowed. "I see," she murmured.

Merlin studied Nimue's face. "You did not ask about the horn."

Nimue was silent.

"It's a horn that will summon-"

"Don't say it!" Nimue yelled.

It was too late. "The Wild Hunt," said Merlin, a twinkle in his eye. "Were you expecting me to say a different name?" he asked, seeing the relief wash over Nimue's face.

Nimue shivered. "I'll come," she said in a quiet voice.

"All these treasures are in my cave," said the Magician. "It is not far from here."

When they stepped out of the lab and into Mendell's garden - Nimue's garden - the Fairy Queen was absent.

Merlin led the way. At Hengoen Hill, Nimue bent down to pull a weed. Merlin grabbed Nimue's hand (the lame one) and lifted her up roughly.

"This way," he said, turning to the West. Nimue pulled her hand away.

"You and I shall make a wonderful team," said the Magician, his voice softer. "Imagine meeting people from other dimensions, from other times, from legends. And not just in dreams, but in flesh and blood. We'll use the jewels and the beds to bring them right here to Camelot!"

Nimue looked down at her feet, but her heart fluttered with possibility.

"You were able to coax a rowan berry into revealing its magic."

"Rowan fruit," corrected Nimue under her breath (she'd spent much time with Mendel since writing her original instructions).

"Do you believe that blood contains such hidden magic as well?"

Nimue thought of Mendell's experiments. His conclusions on dominant and recessive traits could apply to all living organisms. What a comfort it would be to have someone with whom to share her knowledge. She opened her mouth to speak but quickly closed it. *Merlin was not hungry for knowledge, but for power.*

"I know of nothing but peas," Nimue said meekly. A breeze blew strands of red hair into her face.

Merlin took in Nimue's beauty and smiled. "You may remember more when we reach the cave."

By the time they arrived at Merlin's cave, the wind was blowing fiercely. An enormous boulder blocked the entrance. The Wizard pulled a wand from his cloak and tapped the rock. It rolled aside at once. Merlin eagerly pulled Nimue into the cave. Here, the howling wind was silenced.

The bed was as gigantic as Merlin had described and scientific curiosity overtook Nimue once again.

"Is it indeed a giant's bed?" she asked.

Merlin shook his head. "My own creation," he replied, pouring himself a glass of wine.

Nimue ran her fingers across the bed's massive frame. Her eyes locked on a peculiar device which had been affixed to the headboard. Nimue brought her face to the contraption and began inspecting it.

"My own invention," bragged Merlin, finishing his wine and pouring himself another glass.

"Is it a self-contained energy source?" asked Nimue. "How does it work?"

Merlin was already on his third glass of spirits. He mumbled indiscernibly. Nimue removed the device's back panel and began studying its inner workings. The Wizard moved behind her and grabbed the sleeve of her gown. Smelling Merlin's breath on her neck, Nimue spun around. The end of her emerald sleeve tore in his hands.

"This will do," said Merlin, placing the piece of torn fabric on top of the machine. He began turning dials and pressing buttons.

Before long, an image appeared on the bed. Had Nimue lived in different times, she would have recognized the image as a hologram. But, alas she did not. To her, it was but a ghost that had materialized before her eyes. Nimue froze. The image was blurry, but its face was familiar. And its voice - William's.

> *"And blessedness goes where the wind goes,*
> *And when it is gone we are dead;*
> *I see the blessedest soul in the world-"*[39]

Merlin nodded his drunken head. The image of William faded and Nimue, weak with shock and regret, fell to the floor.

"This is only a demonstration," said Merlin, his speech slurred. "Like your friend's peas. Imagine the possibilities when blood turns to flesh!"

Nimue remembered that a drop of William's blood had landed on the sleeve of her gown all those years ago. She gasped. Merlin pulled her up, although he could barely maintain his balance.

"Imagine the possibilities," he repeated. "The beasts we can resurrect! The armies we will command! The Black Pig will rise again-" Before finishing his sentence, Merlin slumped over, landing face first on the bed. His loud snores soon filled the cavern.

Nimue knew the rest of William's poem. She spoke the next verse aloud and felt the strength returning to her body.

> *"O blessedness comes in the night and the day*
> *And whither the wise heart knows;*
> *And one has seen in the redness of wine*

---

[39] William Butler Yeats, The Blessed, 1899

The Incorruptible Rose," [40]

Nimue, deciding that this was not a time to 'dwell in Possibility'[41], grabbed Merlin's staff and ran for the door. *Was it a horn that lay in the corner?* Nimue paused, but a loud snore from Merlin refocused her attention. She ran out of the cave and, without hesitation, tapped the rock with the staff. The staff broke in two as the boulder rolled back against the opening of the cave. The Wizard was entombed.

Nimue let out a deep sigh of relief. The job was done. The wind howled loudly in her ear. So loudly, in fact, that Nimue did not hear the Fairy Queen's approach. In her hands, the Queen held a fishbowl.

Play Along: Solve the Chinese Remainder problem, find out why the Aspen is the largest organism on Earth, and read more Arthurian legends featuring Merlin – all in this chapter's activities. Join the fun on page 245.

---

[40] William Butler Yeats, "*The Blessed*", 1899
[41] Emily Dickinson, "*I dwell in Possibility*"

11. Merlin's Cave

Lynette and Elizabeth arrived at Merlin's "secret cave" with the dawn. There was nothing discrete about the enormous metal door affixed to the face of the mountain. At its center, the letter "M" gloated in the first beams of morning light.

"What happened to the legendary boulder?" Elizabeth asked.

Lynette pointed to an enormous pile of rubble some distance away. "The work of a giant," she whispered. "It seems that Merlin has made a new friend."

Lynette turned back to the massive door and began pulling on its handle. "I need to get in that cave," she said.

"Are you sure? Maybe we should meet up with the rest of the Ladies instead," Elizabeth suggested.

"We don't have time," insisted Lynette. "The fate of Camelot may be resting on the objects in this cave." She ran her fingers along the door's "M" and the beams which surrounded it.

Elizabeth pointed to a plaque at the door's base, and read it aloud:

Move two pillars, and not one more,
To release the Wizard and open the door. [42]

"Why would Merlin leave instructions for opening the door to his secret cave?" asked Elizabeth.

"Why would anyone's secret cave have such a pretentious entrance?" Lynette countered.

Elizabeth studied the door. "I wonder if these metal beams are what Merlin calls 'pillars.'" She pushed on one of the beams, and it slid easily as if on a track. Removing another beam so it could be repositioned proved more challenging. Lynette and Elizabeth finally managed to get the beam dislodged and put back into its new location. Elizabeth stepped back and admired their work. The letter "M" was no longer surrounded. "The Wizard has been released... I think."

Lynette tried the handle once more, and this time, the heavy door swung open. She put a hand on

---

[42] Write the letter M on a piece of paper and use toothpicks for the four pillars of the cave (see the diagram). Move only two toothpicks so that the letter M is completely out of the cave, but the cave retains its shape. The answer is on page 250.

Elizabeth's shoulder. "You'll have to hide behind a tree and be the lookout," she instructed. "If you see danger approaching or anything out of the ordinary, shriek like a Great Horned Owl. If I don't come out in five minutes, take a few steps into the cave and make the clicking and chirping of a swiftlet."

"Swiftlet?"

"Cave-dwelling birds rumored to have been brought here by Merlin."

"Are you sure about this?" asked Elizabeth again. A feeling in her stomach warned her that something was not right. Lynette gave Elizabeth's back a reassuring pat and headed into the cave.

Elizabeth kneeled behind a tree. She kept her eyes fixed on the cave's entrance. *How will I know when five minutes have passed?* she wondered, biting anxiously on a fingernail. Without a watch or clock handy, Elizabeth decided to count to 300. She had reached 245 when Lynette's head appeared from behind the door. She glanced around quickly and then joined Elizabeth.

"A giant has made the cave his home," said Lynette. "His smell is everywhere." She pinched her nose. "Oil lamps were still burning inside. He only recently left, and will likely return any minute. We haven't much time..."

Elizabeth shuddered at the thought of encountering a real giant. Lynette produced a ram's horn from her satchel and handed it to Elizabeth. "We'll take the horn to King Hallden. Who, or what, the horn will summon is unclear, but King Hallden will know what to do. Merlin seemed to think it will call the Wild Hunt. Whatever you do – don't blow it."

The warning was completely unnecessary, as Elizabeth was not about to put her lips to this object. "The Wild Hunt?" she managed to ask.

"An army of huntsmen, some Fairies, others dead spirits."

"Evil things."

"No, not necessarily. Some believe that the horn will wake Arthur from his resting place in Avalon. He'll lead the Wild Hunt to save Camelot in its darkest hour. If this is true, then King Hallden will need the Wild Hunt's help to fend off the invaders."

Elizabeth shook her head. "Let me get this straight," she said. "Merlin the magician was working for the Devil, and a hunting party led by a dead king is a good thing?"

"Yes. I'd love to explain, but we haven't the time. We don't even know if this horn will indeed call the Wild Hunt. It may be but a useless relic. Or perhaps the horn calls someone else. Nimue was not sure, either." Lynette turned and began walking back toward the mouth of the cave.

"Wait!" called Elizabeth. She secretly hoped that delaying Lynette may keep her from reentering the cave altogether. "Who did Nimue believe the horn would call?"

Lynette spun around. "She thought that Merlin had found Borabu, the horn that calls the Fianna to battle. The Fianna were a band of the fiercest warriors and best poets in the whole of Ireland. To join the Fianna, a man would have to fight while buried to his waist in sand. He would recite an epic poem as he fought, never missing a line."

Elizabeth smiled, trying to picture this in her mind.

"As they sang out their mottos, armies would tremble in fear," continued Lynette. "*Purity of our hearts, strength of our limbs, action to match our speech!*"

"The three-petaled way!" exclaimed Elizabeth.

Lynette paused for a moment to reflect on this before continuing. "The leader of the Fianna was none other than Fion Mac Cumhaill. There are many fabulous tales to tell about the Fianna under Fion's command, but I must go back into the cave."

"Why?" asked Elizabeth. "We have the horn. Let's get out of here."

"There's a bed in there," said Lynette excitedly. "It's number 84. This bed and the one on which you and your sister arrived in Camelot are friendly pairs. And its gem is glowing! There's a device of some kind attached to the headboard. I think it's the energy source we've been looking for!"

Once again, Lynette disappeared into the cave, leaving Elizabeth hunched behind the tree. This time, Elizabeth occupied her mind by finding the divisors of 84 and then 270 as she ran her fingers across the horn. After determining that the two numbers were indeed a friendly pair, her thoughts turned to what a friendly pair of traveling beds might mean. Could the giant use the bed to travel to the cabin in Lake Despinassy?

For Elizabeth, the very concept of a giant was akin to the image of Merlin as a cartoon character. She scanned her mind, searching for references to giants she may have encountered in literature. If Elizabeth was well-rested, she might, at a minimum, have remembered the Cyclops from Greek mythology, Grendel from the Beowulf saga, or Ogias from *"The Book of Giants"*. As it was, however, Elizabeth and Lynette had rested for only minutes before descending from the hilltop garden. "Fi,

fie, fo, fum," were the only words that played in
Elizabeth's head. The idea of a giant man fitting into that
tiny cabin cellar almost brought a smile to Elizabeth's
face (but only almost).

In the tree above Elizabeth's head, a nuthatch had
emerged from his tree hole nest. Clinging to the trunk
with his strong feet, he sat still, upside down, watching
the girl. The nuthatch was not interested in passing
judgment like the Pigeon. He was only waiting to see if
the girl would stir some insects from the tree – *breakfast!*

The girl, however, was sitting very still. *Had she
fallen asleep?* Just as the nuthatch was preparing to seek his
provisions elsewhere, the tree began vibrating. He moved
to a higher branch and looked out over the forest canopy.

The nuthatch observed the treetops convulse, one
after another as if swept by a localized gale. As the line of
shaking trees drew near, the forest floor crunched
beneath what could only have been an enormous weight.
The nuthatch tensed. There was now no doubt about it -
a giant was approaching!

A chickadee was the first to ring her warning
"chicka-dee-dee-dee-dee-dee". The nuthatch recognized
the severity of the impending danger by the chickadee's
extra "dees". He added his low siren. "Wha-wha-wha."
Soon an explosion of chirps filled the forest. The girl,
too, added her high-pitched shriek (*a pathetic imitation of a
Great Horned Owl!*)

The giant seemed undisturbed by the forest
cacophony. He paused briefly at the foot of the cave and
looked around. The giant's prominent nostrils flared and
as he sniffed and grunted, the branches of the nuthatch's
tree bent with the force. As the giant man pushed his
way into the cave, the bird retreated deeper into his hole

and a haunting silence fell over the forest like a thick blanket.

Back on the ground, Elizabeth's eyes locked on the mouth of the cave for any sign of Lynette. "Get out! Get out!" she mouthed. The time passed (*was it seconds or minutes or hours?*), the silence lifted, and the sounds of the forest returned.

Something was stirring inside Elizabeth (*could it be courage?*). She recalled how Lynette had swallowed the bitter fruit for her. She had chosen this act of self-sacrifice without being asked, without any hesitation. And so, Elizabeth pushed her instincts of self-preservation to the side. She was ready to help a friend.

Not sure if she was even herself, Elizabeth crawled toward the massive open door of the cave. She took a breath and peered inside. Beyond the door, lay an enormous cavern. Stalactites hung from its ceiling like sharp dripping daggers, and stalagmites reached from the floor like witch's hands. Oil lamps were staggered around the floor of the cavern, and dark shadows flickered on the walls. The walls themselves were alive with the movement of a million nesting birds. *Swiftlets!* A putrid smell hit Elizabeth's nostrils. Was its source the bird dung that littered the floor of the cavern, or was it the smell of the giant?

Elizabeth's eyes were drawn to the center of the room, where she saw the giant bed which appeared to be an identical copy of their own ship-bed. The ruby at the heart of the familiar trefoil symbol cast a fierce red light upon the two dragon heads. Elizabeth thought, at first, that a black cat was sitting on the headboard, but soon realized that this must be the device that had drawn such excitement from Lynette.

Emboldened by the dragon heads, Elizabeth covered her nose with a shirtsleeve and crept into the cavern on silent feet. The sound of the swiftlet's clicking echoed from every crevice of the cave. The shadows flickered around her. One of the shadows, larger than the rest, caught her eye. As Elizabeth watched, the shadow morphed into the form of a giant man. His humongous frame was bending over something. Elizabeth hid behind a towering stalagmite and pressed her body against the moist wall of the cave.

The hollow sound of footsteps approached. The giant came into view with a large wooden box tucked under each arm and a blue jar balancing under his chin. He dropped the boxes onto the bed and placed the jar on the headboard. Elizabeth held her breath as the giant rose to his full height.

His booming voice echoed from every corner of the cavern at once.

> "*A delusional fool in a giant bed lies,*
> *But he's only a giant within his own eyes.*
> *The mirror of dreams will reflect but a man*
> *Who will never accomplish what Ogias can!*"

Ogias chuckled to himself and his great chest heaved. The cavern shook. The giant rubbed his large hands together. The sound was that of rocks striking against each other, and the friction produced a warmth that reached Elizabeth's face.

With one hand in each box, the giant ceremoniously pulled out two pieces of fabric. To Elizabeth, they appeared the size of handkerchiefs, but between Ogias's thick fingers, they were no larger than buttons. He dipped each hand into the jar and immediately dropped the fabric scraps into the 'device'.

129

Ogias continued singing:

"The blood of an Englishman makes a fine feast,
But it pales to the powerful blood of a Beast.
With a Fee and a Fie and a Foe and a Fum
In that sleep of death, what dreams may come?"

Elizabeth decided that although the giant was no William Shakespeare[43], he did have a certain way with words.

The covers on the bed began to move. They rose and ebbed higher and higher as if a giant serpent lay beneath them. Elizabeth stared, mesmerized, as Ogias pulled back the bedspread to reveal two monstrous figures tangled in each other.

"Welcome, Grendel," he addressed the smaller of the two, pulling him up by the forearm. Grendel's other arm was missing. His wide grin revealed a mouth of sharp teeth.

The other figure continued to struggle with an unknown enemy on the bed.

"Who's there?" he yelled out.

"Nobody," replied Ogias with a chuckle. The words sent the third giant into a rage. He flew off the bed, clawing wildly at the air.

Ogias placed a broad hand on his shoulder and pushed him down. "Polyphemus, my friend," he said, "it was only a joke."[44]

---

[43] *"For in that sleep of death what dreams may come"* is a line from *"Hamlet"*, a play by English playwright William Shakespeare (1564-1616).

[44] If you are not familiar with the story of Odysseus and the cyclops, you may wish to read the summary on page 254 to get the "joke".

Elizabeth noticed that this third giant had a gaping hole in the middle of his forehead instead of an eye.

"The cyclops," she whispered.

"Who said that?" the cyclops yelled out, bolting up from the bed. Grendel spun around and snarled.

"Said what?" asked Ogias, pushing Polyphemus back down again. "Let's have a look at that eye." He leaned over the cyclops and placed a hand on each side of his face. "Merlin and the Black Pig will be making their move soon. We must prepare for battle."

It was then that Elizabeth saw Lynette peer out from beneath the bed. Her face, flanked by Ogias's legs, was pale as the moon. She looked Elizabeth straight in the eyes. It was a look of desperation. "Get help," she mouthed.

Elizabeth's voice was not her own. "*I'm Nobody! Who are you? Are you – Nobody – too?*"[45]

The Cyclops flew back to his feet again. The two giants' heads collided with a loud, dull thump.

Elizabeth did not wait to see what happened next. She rose to her feet and bolted out of the cave and into the woods. The thought of waiting for Lynette crossed her mind only momentarily. The goal of the distraction was to given Lynette an opportunity to escape. But something - heart, head, instinct - told Elizabeth to keep running, not to look back.

Brambles cut at Elizabeth's legs, but still, she ran. The cold air pierced her lungs, but her pace never slowed. The aspen leaves no longer whispered - they snapped and screamed – but Elizabeth's determination never wavered.

---

[45] Emily Dickinson, "*I'm Nobody! Who are you?*"

Seconds and minutes flew by as Elizabeth fled. Soon the passage of time grew into hours. *Hours!* Her legs were sore and weak and tired, but stopping was not an option – it was never an option.

> *"If you can force your heart and nerve and sinew*
> *To serve your turn long after they are gone,*
> *And so hold on when there is nothing in you*
> *Except the Will which says to them: 'Hold on! '"*[46]
> -Rudyard Kipling

Clarity, brought about by a heightened state of danger, fought valiantly against fatigue. "I must get help," Elizabeth said aloud, clutching tightly to the horn. Her voice was drowned out by the wind. "Get help. Get help," she repeated.

Her stomach gave a sudden jump. It was a familiar feeling, as Elizabeth was accustomed to beginning each night's sleep with a tumble that would startle her back into reality. She was not one to submit to the irrational world of dreams so easily. A lucid dreamer, Elizabeth would narrate her dreams with scientific explanations. But now - now that she had seen giants and magic at work - she allowed herself to succumb to... *was it faith?*

Whether the path of flickering lights she saw was generated by chemicals in her brain or by Fairies, Elizabeth did not know or care. It was a path of light, and a direction to follow is always better than a random walk algorithm (a blind stumble). Following the Fairy-path, Elizabeth crawled into a clearing. "Get help," she whispered before collapsing from exhaustion.

---

[46] Rudyard Kipling, "*If—*", 1895

When Elizabeth awoke, the wind had ceased entirely. She sat up, disturbed by the unnatural silence. Silence, too, needed a soundtrack – a poem. Elizabeth strained to remember the words of William Butler Yeats in Lynette's story (Nimue's story). "*And blessedness goes where the wind goes, and when it is gone-*"[47] Elizabeth shuddered.

The clearing was bathed in sunlight and at its center stood... a bed! Elizabeth suddenly felt not only awake but alive. She stumbled towards the bed and ran her hands across the flowered quilt. This was the bed upon which Lulu was sitting when she last saw her.

Elizabeth looked at the metallic flower decorations on the headboard, some open and others closed. She skimmed the instructions: "Base two," Elizabeth murmured, feeling the embroidered flowers between her fingers.

The jewels must have faded before Lulu and the Ladies reached Victoria, Elizabeth concluded. She reached into her pocket and pulled out the two new amethysts. They were still glowing "with the energy of youth" (as Lynette had said). Returning one to her pocket, Elizabeth began working on removing the faded jewel from the large flower decoration on the headboard.

Without a tool, Elizabeth was forced to pry the metal flower open with her bare hands. Her fingertips were bleeding by the time she had finished. Elizabeth pressed the shining jewel into its new home using the palm of her hand. She had to push forcefully, bracing both legs against the headboard, but finally the jewel clicked into place. It sparkled brighter than ever.

---

[47] William Butler Yeats, The Blessed, 1899

"Now for the part I dread," said Elizabeth aloud. She was talking about falling asleep, of course. But something didn't feel right. Elizabeth clenched her fists. *Should I check Lulu's calculations?* she wondered. It would give me something to do. Elizabeth began straightening out the flowered quilt. At the sight of her hand, she paused and laughed. Uncurling her fist, Elizabeth revealed a perfect rose in the center of her palm.

"I didn't think I was that impressionable!" said Elizabeth. She was too fatigued to laugh at her own joke. Besides, her friend was in danger. She turned to the small flowers on the headboard: open, closed, open, closed, open, closed, open. Elizabeth read the clue again, then looked at her palm. The old Elizabeth - a rational Elizabeth - would have said that the impression of a rose on her hand would have zero relevance to the answer. But this Elizabeth - the new Elizabeth - had seen giants. She took another look at her palm. "Now to add five in binary..."

Play Along: In this chapter's "play along" section, you will solve toothpick puzzles, learn to identify friendly numbers, and read the back-stories of Ogias, Grendel, and the Cyclops. Turn to page 250 to join the fun.

12. Nimue Anew

Lulu stood at the edge of the Pool of Joy feeling despondent. She hugged the fishbowl to her chest with one arm. The other was weighed down by the Fairy Queen's crown. Lulu remembered how she had crouched in darkness inside the tree as the tempest tore at the pine like a raging giant. The sounds of howling wind and cracking wood still rang in her ears through the silence and stillness. "More ephemeral than a flower," she whispered to the fish. "Such is the transient madness of a storm." The lion statue nodded its great head.

The crown was heavy in Lulu's hands. After the storm had passed, Lulu returned to the balcony to look for the Fairy Queen. She found only her crown, glistening in the first rays of morning. "All these lands could be yours," the Queen had said. Hours ago, Lulu would have jumped at the opportunity. Now that she had felt the weight of the unwieldy burden - the crown - in her own hands, she was not so sure.

When Mrs. Magpie had mentioned that she would need training to become Queen of Tulgey Wood, Lulu had foolishly imagined classes in math and perhaps even etiquette. Instead, she had received lessons in character. Lulu put the crown down on the ground and placed the small fishbowl in its center. The Fish of Desire swam in circles. "Here's your crown back, Nimue," said Lulu. "If that's who you are." The fish's eyes twinkled.

Lulu found the two basins, which the storm had moved against the trunk of the tree. She lifted the larger basin up by its chain. Lulu had an important task to complete. She filled, poured, and emptied the hazel water between the pool and the two containers until the larger basin held exactly four units of the magic water. Lulu carefully removed this basin from its chain and carried it to the bed which sat next to the pool, surprisingly dry and undisturbed. All its pins and marbles were in perfect order.

She then returned to retrieve the fishbowl. For a moment, Lulu wondered if she should leave the crown behind. She ultimately decided to bring it, however, as the crown would be far too dangerous in the wrong hands. Lulu turned the switch on the bed from 'THERE' to 'HERE'.

A fish's dreams, a fairy's trail,
A giant's screams, the Black Horse Vale.

Lulu blinked and sat up. Lady Olwyn and Lady Elfinhart were standing in front of the bed. Each Lady posed with, not three fingers, but an open hand pressed against her heart. The Ladies' dresses were now crimson, and the badges had changed from lilies to roses. "Your clothes-" Lulu began.

"The news came by Populus Post just minutes ago. Camelot will soon be at war," replied Lady Elfinhart. "You may no longer address us as Lily Maidens, but as The Incorruptible Roses of the House of Orkney and the Sisterhood of the Traveling Beds, defenders of honor and the five-sepaled way."

Lulu lifted an eyebrow. "Isn't a rose, you know, a dicot?" she asked in a low voice.

"We still defend a single code of honor, if that's what you're asking," Lady Elfinhart replied sharply. "One must not change one's core values in changing circumstances, you know. We still consider ourselves Lily Maidens on the inside. But on the outside - we're ready to use our thorns!"

"From *hence* the defense of tenderness!" added Lady Olwyn, dramatically pulling a sword from her belt.

The lily and the rose. Where have I seen that before? Lulu wondered.

"Will you go into battle?" she asked, feeling for the Vorpal Blade against her chest. She had entirely forgotten its presence until just a moment ago.

"Our primary objective is to protect the beds at all cost. They are our gateway to knowledge. Imagine if someone had saved the Library of Alexandria from burning..."

"The beds! My mission!" Lulu suddenly remembered the hazel water. She had a job to do. With the large basin in her arms and a look of determination on her face, Lulu headed toward the door of the cabin. The Ladies moved aside as she passed.

The wind blew gently across the Lake of Tears, sending sparkling ripples across its surface. Looking closer, Lulu saw that the water was dotted with jewels that bobbed around the lake like lost buoys.

Not long ago, Lulu had imagined herself at the center of a grand ceremony, singlehandedly diluting sorrow and saving the jewels to the sound of cheers and accolades. After meeting the Fairy Queen, however, things were different – Lulu was different. She silently

137

emptied the basin into the lake. The jewels winked in gratitude (*or was it farewell?*) as they sank beneath the water.

Lulu, light as a bird, spun in a pirouette. She suddenly noticed Victoria sitting patiently at the water's edge. The water lily's blossom had changed from white to a dark reddish-purple. "Are you also preparing for battle?" Lulu asked Victoria.

"She has indeed changed her attire early," came Lady Elfinhart's voice. Lulu turned around to see the two Ladies standing stoically behind her. *How had they approached so silently?* Lady Olwyn was holding the Fairy Queen's crown, and Lady Elfinhart cradled the fishbowl.

"Usually the transformation happens on the second night," said Lady Elfinhart. "Perhaps this is a day for new beginnings. Watch!" She pointed to Victoria.

The crimson flower opened, and the black beetle appeared. At first, he wobbled around the flower, disoriented. As Lulu watched, the beetle regained its bearings and took flight, a trail of pollen glittering behind him.

"The beetle may take his part in the battle later," Lady Elfinhart continued. "But first, he has a promise to keep - another Victoria to find."

"There are other Victorias?"

"Of course," replied Lady Elfinhart. "She will never feel the sadness of being a singularity. This one is no longer a Victoria, by the way. Victor would be a more appropriate name."

"She- I mean 'he'- has changed genders?" asked Lulu, confused.

Lady Olwyn's voice rang out clearly.

> *"What now is a Victor was once a Victoria,*
> *Is it botany, politics, or phantasmagoria?"*

"It's botany," said Lady Elfinhart. "Definitely botany."

Lulu was not quite sure that phantasmagoria did not play a role in the affair. In this strange dream-like world, nothing was ever as it seemed. She took another long contemplative look across the lake.

Lady Elfinhart cleared her throat. "Now to the next order of business. "Where is the Fairy Queen?" She was staring at the crown in Lady Olwyn's hands.

"I- I don't know," fumbled Lulu. "She made a wish and disappeared. I saw it with my own eyes."

"In that case, congratulations. The crown is yours. It is the custom that the last being to set eyes on a Fairy Queen when she leaves her post must take her place." Lady Elfinhart delivered the news without emotion. *Was she announcing a prize or a sentence?*

Lady Olwyn held out the crown. The old Lulu would have jumped at the opportunity of becoming a Fairy Queen. The old Lulu would have already been planning her acceptance speech. But this Lulu - the new Lulu - had felt the weight of the Fairy crown. This Lulu had felt the weight of wishes. She hesitated.

"I can't," said Lulu finally "I'm not even a Fairy."

"Easily remedied with a drink of hazel water," Lady Elfinhart reassured her. She held the fishbowl out toward Lulu who shook her head adamantly and took a step back from the offering. She longed to have her sister by her side.

Lady Elfinhart took a large step forward and pressed the fishbowl to Lulu's chest. Lulu was now pinned between Lady Elfinhart and the edge of the lake. She smelled the Lady's flowery breath. The fish in the bowl leaped, and Lulu nearly tumbled into the lake as she bent back to avoid the splash.

"Wait!" Lulu yelled out. "There was another witness - the Fish of Desire." She flipped around and tossed the entire bowl into the water. Lulu put a hand over her mouth wondering what she had done. Back at the Pool of Joy, she had worked so carefully getting exactly four units of hazel water. *Will the extra amount from the bowl disturb the balance of the gems? Can tears be so diluted that those who forget sadness are doomed to repeat the mistakes of the past?* There was no time to analyze such matters.

The lake darkened to the color of tea, and tiny bubbles rose to its surface. A head popped out of the water at the far shore.

"Nimue?" Lulu called out "Is it -," but before she could finish her thought, another head emerged- this one much closer. The head looked at her with large child-like eyes. Soon another popped up beside it.

"Fairies," Lady Elfinhart whispered.

"The world is full of weeping," one of the Fairies sang out. "So very full of weeping," the others echoed.

This time, Lulu felt she understood. Tears streamed down her face. "So very full of weeping," she cried. *Was exhaustion responsible for this sudden release of emotion?* Lulu stepped back, remembering the hands that had reached for her when she'd crossed the lake aboard Victoria. *Had it only been yesterday?* The Fairies took no interest in Lulu, however. They were staring intently at a spot directly between the Lake Isle of Innisfree and the nearest shore. In this location, the water bubbled wildly.

Lulu felt Lady Elfinhart's hands on her shoulders, and now a new emotion stirred. The fairy song was no longer audible. It had been drowned out by a jubilant cacophony of birdsong.

Thousands of birds alighted on the trees surrounding the lake, and each was singing in exaltation, each song in a unique key. Birds treat sadness as they do fear - as naught but a fleeting emotion, a protective instinct. For the birds, happiness is the default state of being.

> *"If you can dream - and not make dreams your master;*
> *If you can think - and not make thoughts your aim;*
> *If you can meet with Triumph and Disaster*
> *And treat those two impostors just the same;"*[48]
>
> -Rudyard Kipling

Birds never mope. Birds never gloat. Birds never hold a grudge. (Such is the case for the birds of Earth and Camelot, at least, as Wonderland birds are very different). Lulu listened to their song and found herself laughing.

A deep rumbling from far beneath the lake provided a bass to the birdsong. As Lulu watched, smiling, the spot that was bubbling more fiercely than the rest began spurting and surging higher and higher until it erupted into a great fountain. Birds and Fairies grew silent, and everyone listened to the splashing, tumbling water.

> *When your soul holds joy like a fountain,*
> *A single drop can move a mountain.*

Lulu tensed up as the fountain turned blood-red, but a glance at Lady Olwyn's face let her know there was nothing to fear. The fountain stopped spraying, and the figure of a woman appeared in its center, her hair as red as the receding water. The last of the fountain's spray landed with a splash at Nimue's dainty feet, which stood upon a new water lily.

---

[48] Rudyard Kipling, "If—", 1895.

Nimue floated across the lake upon the Victoria, one arm raised and the other lying limply at her side. The former Lady of the Lake, who only minutes ago had been the Fish of Desire and would shortly become the new Fairy Queen, was floating straight toward Lulu.

She was a vision of beauty nonpareil! Nimue's silver dress sparkled like reflections upon rippling water, and her sleeves were an exquisite shade of green, rich and deep. But Nimue was no longer William's Glittering Girl, but a woman, wearing the creases of time and wisdom proudly in the corners of her eyes and mouth. The new Victoria slowed to a gentle stop directly in front of Lulu.

The Fairies in the water reached for Nimue with their child-like hands. She knelt for a moment to plant a tender kiss upon each little head, but the hand for which she reached was Lulu's. Lulu fell to her knees and bowed her head as Nimue stepped onto the shore, her small white feet a stark contrast to the dark sod of the bank.

"Look at all this fuss," said Nimue shyly. "It is quite unnecessary."

"It is the custom," replied Lady Elfinhart. Lulu raised her head and saw that the Ladies, too, were kneeling before Nimue.

Lady Elfinhart rose and nodded to Lady Olwyn who stepped forward and placed the crown upon Nimue's head.

> *"Golden Rowan of Menalowan*
> *was crowned the Fairy Queen.*
> *Her allegiances firm, her heart was pure,*
> *her sleeves were colored green.*
> *Golden Rowan of Menalowan*
> *had found what she had sought;*
> *Blessedness, beauty, and truth she'd bring*

to the whole of Camelot!"

A harmonious song of glee arose from birds and Fairies, alike, and Nimue blushed. The newly-crowned Fairy Queen listened appreciatively, a pleasant smile on her face. Then her look turned somber.

"There will be a time for fanfare," the new Fairy Queen promised. "But that time is not the present. Dark shadows loom over Camelot and war will soon rear its ugly head upon our lands. In the past, the Fairy Folk have been hesitant to interfere in human affairs. Sometimes inaction is akin to the wrong action. We must fight in the name of goodness. We must fight for what we know is right. We must prepare for battle."

The most sensible thing I've heard since we arrived in Camelot, Lulu decided. Her thoughts turned to her sister.

The fairies nodded in understanding and disappeared beneath the waters. The birds scattered. The aspens shuddered and whispered.

"What do you order us to do, Your Majesty?" asked Lady Elfinhart. Nimue appeared startled at being addressed in this manner. *Was she adjusting to the title of her new post, or was it that human allegiances lay with human monarchs, not Fairies?*

Lady Elfinhart saw this too. "Nimue," she said, her voice gentle and comforting. "You are one of us. We know you and trust you. And you, in turn, can count on the Lily Maidens-" Here, Lady Olwyn gave her a nudge. "-I mean the Incorruptible Roses," Lady Elfinhart corrected herself. "We defend a single truth and will do whatever it takes to help King Hallden."

At the mention of King Hallden, a look of determination appeared on Queen Nimue's face. "Bring me the Fairy Harp," she ordered.

143

Lady Olwyn ran toward the cabin and promptly returned with the instrument. The wood of the harp was splintered, and the strings were torn.

"The Fairy Queen – I mean, the *former* Fairy Queen - must have destroyed it," said Lady Elfinhart, searching for an explanation.

"I was there," said Nimue, "remember?" and it was Lady Elfinhart's turn to blush. "It is of no great consequence. Lulu - the basin, please. Fill it with lake water."

Lulu searched the shore for the location where she had dropped the water basin and saw it immediately. Filling the large container with lake water without falling in was tricky. Lulu managed the task, finally placing the full vessel at Nimue's feet. Lulu stared in awe as Queen Nimue produced a ruby from the folds of her garment and dropped it into the water. Nimue then sat down on the grass and began removing the strings from the battered harp. Holding a string in her delicate fingers, Nimue strung one end through the loop from which the metal chain had hung. She stretched the string across the mouth of the basin and secured the other end in a groove on the opposite side. The string passed through the basin's center. Nimue plucked the string, and as it vibrated, a hauntingly beautiful tone filled the air.

"As Pythagoras discovered, the magic of music, whether produced by lyre, harp, or basin, lies not in the tone, but in the interval," she explained. "And the first musical interval we need is the perfect fifth." She handed Lulu another harp string. "Lulu, will you kindly stretch this string so it is two-thirds the length of the first?" Lulu nodded excitedly and began measuring the first string with her fingers.

"Precision is critical," Nimue warned. "In the folly of my youth, I once strung the harp in haste at a fairy wedding. I played a minor third by mistake. To you, it would sound like the first two notes of '*Brahm's Lullaby*'-"

"Or the start of '*Greensleeves*,'" whispered Lady Elfinhart under her breath. Perhaps she was upset that Lulu was being assigned the important tasks.

Queen Nimue retained her composure and was only betrayed by a barely-audible sigh.

"What happened next?" pressed Lulu who was now using a long blade of grass to make the measurements.

"That musical interval sent the entire wedding party into a deep sleep which lasted for three days," said Nimue with a laugh that sounded like tinkling bells. The smile on her face, however, held no joy.

Lulu had folded the blade of grass into thirds and was now fastening the end of the string in the prescribed notch.

Queen Nimue plucked the two strings in succession. It sounded like the first two notes of '*Here Comes the Bride*' (or was it '*Amazing Grace*'?).

As the sound faded and the water in the basin cleared, the image of a young man came into view. His features were handsome, and the look on his face was serious and honest. Although the man wore no crown nor royal robes, his identity was evident.

"King Hallden!" Lulu gasped. Nimue nodded.

A group of young men was gathered around him. The Forest Primeval rose dark behind them, its moss-laden trees standing stoically at attention. These bearded wise men, however, provided no counsel. The young king was alone to make some of the most important decisions of his life.

"Can we hear what he's saying?" Lulu asked.

Nimue gathered a handful of acorn caps and placed them along the indentations of the water basin. Lulu had to strain, but soon she could hear King Hallden's words clearly.

"-and so, now you have been informed of all the same news that I have received via Populus Post," King Hallden told his men. The resolve in the young King's words matched the look of tenacity upon his face. "My enemies would say that such information should be withheld from mere soldiers. They would call the transparency of my heart and mind a weakness. But I believe that a man who fights out of obligation or fear and a man who fights with a thorough understanding of what is truly at stake, are not one and the same. And so, I ask you - my men, my friends- to prepare yourselves for the long fight ahead."

Lulu looked over at Queen Nimue who was nodding in approval. Lulu now understood that this business with the basin was not a silly distraction, but an important way for the new Fairy Queen to see what sort of man King Hallden was before entangling her Fairy Folk in a war that would inevitably cost lives.

Nimue turned to Lady Elfinhart. "Send the message to King Hallden by Populus Post that the new Fairy Queen pledges the commitment of the Fairy Army to Camelot's plight." Lady Elfinhart was obviously excited to have a critical mission. She jumped up at once and dashed toward Victor.

"I can help-" began Lulu, curious to learn more about the mechanics of the elusive Populus Post.

Nimue looked at Lady Elfinhart. "Take Lady Olwyn with you," she instructed.

Nimue turned back to Lulu. "I'll need your help here."

The two Ladies boarded Victor. Lulu turned her attention back to the basin. The men had gathered in a tighter circle with each one resting an arm on another's shoulder. They were singing together, their voices low and solemn. King Hallden's clear voice rose above the rest:

> *"So, we'll go no more a roving*
> *So late into the night,*
> *Though the heart be still as loving,*
> *And the moon be still as bright.*
>
> *For the sword outwears its sheath,*
> *And the soul wears out the breast,*
> *And the heart must pause to breathe,*
> *And love itself have rest.*
>
> *Though the night was made for loving,*
> *And the day returns too soon,*
> *Yet we'll go no more a roving*
> *By the light of the moon."*[49]

The voices faded and so did the image in the water. "I know why their song was a little sad," Lulu blurted out. Queen Nimue looked at her. "Growing up is not always fun," said Lulu quietly. She wanted to elaborate further but decided to leave it at that.

Nimue lowered her head in understanding. She handed Lulu another string. "The next musical interval is

[49] (George Gordon) Lord Byron, "So We'll Go No More a Roving", 1830

a perfect fourth. The string lengths should be in a ratio of four to three." Lulu worked quickly, measuring the first string on a new blade of grass then folding it into fourths. Soon, the job was completed, and the third string was secured in its place.

Nimue plucked the first string and then the new one. These notes sounded like the start of '*Twinkle, Twinkle Little Star*' (*or was it the "Star Wars" theme?*). The water in the basin turned dark as night. As its surface smoothed, a sea of black figures appeared. Whether they were men or beasts was not clear. The dark mass opened and a black knight in full armor rode into view.

"The Black Pig," Nimue whispered.

A man in a long charcoal robe positioned himself next to the black knight. He waved his staff in the air and appeared to be yelling. Nimue gulped and turned pale. "The Magician! Merlin!" She adjusted the acorn lids, and the sound buzzed:

> *'Perfidy and treachery,*
> *Faithlessness, and spite.*
> *The Devil joins the revelry,*
> *And blood will run tonight!"*

The Black Pig began to remove his helmet. Lulu caught a glimpse of sharp fangs before the image disappeared.

"Can we get the picture back?" Lulu asked frantically.

Queen Nimue plucked the strings of the perfect fourth again. The clear tones of the interval rang out, but the image did not return. Nimue looked Lulu in the eyes. "Merlin's magic is too powerful. We have seen enough." She turned away, her gaze distant. Lulu, too, looked across the lake as she wondered what part she would play in the battles ahead. That's when she caught first sight of the figures floating toward her. There were not two figures atop Victor, but three. As the water lily moved close enough for Lulu to get a clear look at its occupants, she began jumping up and down in excitement. The "freeform dance of ecstasy" followed, and Lulu did not care in the least whether she resembled a mercurial squab. *Elizabeth had returned!*

Elizabeth was just as excited to be reunited with her sister, but first, she walked toward Queen Nimue and with a low bow handed her the horn that Lady Lynette had retrieved from Merlin's cave. Nimue's hands shook as she received the gift. Elizabeth was then promptly tackled by her sister's embrace. Tears were flowing freely down the twins' cheeks in a reunion that required no words. The Ladies smiled as they watched.

Elizabeth had been caught up in the rejoicing but soon pulled away from Lulu's tight grasp. "Lynette needs our help," she called out. "She's been captured by the giant named Ogias. We need to go at once."

"How many petals has a lily when one is torn away?" said Lady Elfinhart dramatically.

Lady Olwyn quickly jumped in, holding a hand over her heart.

> "*Can a modest rose survive*
> *With four sepals instead of five?*
> *Ipomoea's glory snatched-*"

"There are times when poetry is appropriate, but this is not one of them," Queen Nimue interrupted. "Go save your friend!" Elizabeth decided that this was the most sensible thing she'd heard since they'd arrived in Camelot.

"Take N. Wake with you," added Nimue. Before Lulu could ask who N. Wake might be, a little green mouse emerged from the Fairy Queen's sleeve and jumped onto Elizabeth's shirt. Lulu quickly transferred the mouse to her own shoulder before Elizabeth had time to scream.

Play Along: Explore the mathematics of musical intervals by making your own string instrument in this chapter's "play along" section on page 256.

13. Giants and Riddles

The door to Merlin's Cave was still ajar, and N. Wake entered it without hesitation. The decision to switch allegiances had not been a difficult one for the Mouse – not after Q revealed her true intentions. N. Wake was now determined to prove her dedication to the Ladies and to use her extraordinary spying abilities to aid the right cause.

N. Wake surveyed her surroundings. The layout of the cave matched Elizabeth's description in all but one detail. The giants had moved an enormous wooden table in front of the bed. Rather than sitting on the bed, however, the giants stood at the opposite side of the table, with their backs to the Mouse. They were engaged in a heated argument, all three of them speaking at once.

It was easy for N. Wake to immediately identify the giant who called himself Ogias, as he was the largest, towering at least a head above the other two. The giant to his right was beating Ogias's broad back with a log - no, it was a wooden arm attached crudely to his shoulder with a series of rusty screws and bolts. N. Wake inferred that this one was Grendel.

The wooden arm dealt Ogias a series of heavy blows. Whenever the arm made contact with Ogias's back, the giant would reach behind him and brush it aside as if waving off a pesky mosquito. His other hand was swatting at the third giant who was yelling into his left ear. This one was Polyphemus the Cyclops, N. Wake deduced.

Ogias was the epitome of composure as he allowed his comrades' unruly behavior to continue for several minutes. Finally, his patience worn thin, Ogias let out a deafening roar – "ENOUGH!" and with a push of each hand sent both of the other giants crashing to the floor.

N. Wake saw her chance and quickly scampered against the wall to the back of the cave. It did not take her long to find Lady Lynette who was sitting in a large iron birdcage looking herself very much like a Cockatiel. Disheveled blond locks formed a crest atop her head, and her cheeks were red from rubbing away tears.

At the sight of the green Mouse, Lynette smiled and pointed enthusiastically to the large lock by which the cage was fastened. *A simple lock*, scoffed N. Wake, lock-picking expert. Changing over to the side of "good" had not yet had the proper effect on her humility, and the Mouse secretly hoped there would be a challenge where she could demonstrate her intellectual prowess.

The swiftlets' clicks, the giants' yells, and the occasional clatter of metal or wood as it bounced off the cavern floor drowned out the small tick that came from the lock and the louder groan that arose from the rusty door of the cage as it opened.

Lynette was halfway to the cave entrance when she looked up to see all three giants on their hands and knees prodding the ground in search of something. She flattened herself to the floor and froze. "Here it is!" yelled Ogias reaching for a black jewel - obsidian - that was lying on the floor. He froze for a moment and began sniffing the air.

The giant's eyes fell straight on Lynette. With a single movement, he snatched the woman by the waist, rose to his knees, and dangled her upside-down in one hand. "What have we got here?" he laughed. "A little mousy trying to creep away!" Lynette tried to wiggle free, but the giant's grip only tightened.

N. Wake's heart sank as she bolted for the exit. Lulu was the first to spot the green Mouse. She immediately recognized the look on the Mouse's face as the look of one who has not only failed but failed momentously. The Ladies and Lulu exchanged glances. With a brave yell from Lady Elfinhart, they stormed the cave. Lulu had drawn her dagger - the Vorpal Blade. Elizabeth wasn't quite sure what storming anything entailed, but she, too, ran into the cave yelling and waving her arms along with the others.

The giants looked up, and all three broke into deep, rolling laughter. They seemed pleased to have a respite from the argument. Ogias dropped Lynette onto the bed and held his jiggling belly. As Lynette scrambled to escape, he wrapped his hand around her waist and, still laughing, forced her down beside him. Grendel sat on Lynette's other side. The Cyclops fiddled with his mechanical eye – the obsidian in its center – as he, too, laughed and dropped onto the bed.

The bed shook wildly with their continued laughter and Lynette was in grave danger of being squashed between the giants. Lulu noticed now that Lynette was no longer in the white attire of the Lily Maidens but instead wore the dark pink outfit of the Incorruptible Roses, complete with the rose patch. *When did she have time to change?* Lulu wondered.

153

Lulu and Elizabeth exchanged confused looks. Lady Elfinhart and Lady Olwyn were standing, swords dropped to their sides, and staring blankly at the laughing giants. N. Wake was nowhere in sight.

Lulu wished that they had formed a plan beyond 'storming the cave'. She tried to remember everything the Ladies had told her about giants. Lady Elfinhart had said that giants love games and puzzles. They have a particular weakness for both games of chance and riddles, although they are not particularly good at solving puzzles themselves, especially ones involving calculations. Lady Olwyn had added that giants have a certain way with words and enjoy composing songs and poems (of questionable merit).

Lulu waited impatiently for the giants' laughter to subside. Then she took a deep breath and stepped forward. "I challenge you to a game - a game of minds - to earn back our friend," she said, making sure her voice was loud enough to reach the giants' ears.

The giants seemed pleased. Grendel cheered and rapped his wooden arm on the table. Polyphemus's face twisted into a sharp-toothed grin. Ogias sat up straighter and rubbed his enormous palms together. "Boys, I do believe these little ladies are what we've been waiting for," he chuckled. "Polyphemus - a chair." The Cyclops rose grudgingly from the bed and rolled a large log in front of Lulu. She climbed up. Standing on her toes, Lulu had a good view of the tabletop. Ogias straightened out a piece of paper, which Lulu at once recognized as a map.

"The first challenge, should you choose to accept it -" said the giant with a wink, "is to divide the territories on this map between the three of us." Lulu was already nodding. "There is only one little condition," Ogias continued. "A giant may not rule two territories that share a border." Grendel placed his normal hand on Ogias's shoulders, and Polyphemus's grin widened.

Ogias looked straight at Lulu. "Do you accept this challenge?" he asked.

"I accept your challenge," responded Lulu bravely (but perhaps a bit too quickly).

Ogias pushed the map toward her and flattened the paper with his palm. Lulu recognized the map of England. The map was already divided into regions by crude black lines and various shapes scribbled in charcoal. The castle of Camelot lay in one of these areas.

Before Lulu had a chance to recognize how out-of-place it was to have modern borders of England on what was presumably a 5th-century map, Ogias dropped some wooden coins onto the table. Lulu saw that the red ones bore the likeness of Ogias himself, while Grendel's face was etched onto the green coins and Polyphemus's on the blue.

"The Black Pig awaits us at the Black Horse Vale," explained Ogias. "His vast army will trample Camelot tomorrow. The Black Pig has promised us all these lands following King Hallden's demise." He placed a red coin upon the map - straight on the castle.

"We will meet him shortly to solidify the battle plans," added Grendel.

"But I won't agree to anything until this matter is settled," said the cyclops, raising his voice.

"None of us will agree to anything until the matter is resolved," said Ogias. He turned to Lulu. "I would highly recommend that you begin."

Lulu placed another coin - a green one - on the map. Grendel leaned in to have a closer look.

Suddenly Lynette stood up on the bed and began "singing" in a series of loud whistles and clicks. *Was she calling the swiftlets or was it a secret code?* Lulu wondered. The giants turned to face Lynette, but before they could do anything to her, she stopped singing, smiled and sat back down, her hands in her lap. The giants turned back to the map. Lynette looked at Lulu and winked. What Lulu couldn't see was that Lynette's brief distraction gave Lady Elfinhart just enough time to sneak out of the cave. She was now running through the Forest Primeval, ready to send King Hallden a message by Populus Post with the latest information she'd learned from the giants.

All eyes were once again on Lulu. She turned the coin in her hand over and over. It was a blue one, and Cyclops was holding his breath. *Where have I seen this type of puzzle before?* she thought. "Map coloring!" Lulu said aloud. Her heart sank. "But some maps can be colored with fewer-" she tried to reassure herself. Lulu began frantically placing the coins, one in each territory, then switching them around so coins of the same color would not be placed in adjacent areas.

Mesmerized by the rapidly moving coins, all three giants now leaned forward excitedly with their elbows on the table. Lulu stopped and reflected. She swept her arm across the map, sending the coins scattering. The giants let out a unified growl and appeared ready to jump across the table and devour her right there.

"I'm sorry," said Lulu hanging her head. "It can't be done. The 'four color theorem' states that four colors would be sufficient to color any map so that two territories of the same color do not share a border. Some maps require fewer colors, but this one needs at least four."

Grendel thumped his wooden arm on the table, sending more coins scattering. Polyphemus bared his teeth and would have leaped across the table if Ogias hadn't held him back.

Ogias grunted. "So you're telling us you can't complete the task."

"Because it's impossible," said Lulu looking the giant straight in the eyes, her confidence returning.

Ogias lifted an eyebrow. "You accepted the challenge and failed."

"But the Four Color Theorem -" Lulu began.

"A theory is just a guess," Ogias interrupted, his eyes sparkling. "It's not necessarily true." A broad smile spread across his face. Lady Elfinhart had failed to inform Lulu that certain giants enjoy a healthy debate almost as much as any game or riddle.

Lulu wondered whether Ogias could appreciate the rigor involved in the scientific method. She ultimately decided, however, to forgo an explanation.

"The task is impossible," Lulu repeated. "Nobody could do it." At the mention of his long-time nemesis ("Nobody"), the Cyclops jumped to his feet and overturned the table with a roar. Lulu dove out of her seat and Elizabeth pulled her out of the way of the enormous table just as it came smashing down. The sisters held hands and ran out of the cave. Lady Olwyn was not far behind them. All three stumbled toward the woods, blinded by the bright light of day.

Hiding behind a tree, they looked toward the mouth of the cave, but the giants had not followed. Neither had Lady Lynette. "We have to go back in the cave and save her," said Elizabeth without hesitation.

Lulu agreed. "I'll ask Ogias if I can solve another riddle in exchange for her freedom."

Lady Olwyn looked at them with large eyes and nodded her head enthusiastically. Lulu considered Lady Olwyn for a moment. It struck her, then, for the first time, that the Lady had placed a significant burden upon herself by only speaking in rhymes. Lulu applauded the restraint required for such an approach to life and could name a dozen acquaintances who would benefit greatly from an obligation to select each word so carefully before allowing it to escape from their lips. But such a manner of speaking was cumbersome and clumsy in the face of urgency.

As if in response to Lulu's thoughts, Lady Olwyn's beautiful voice filled the air:

> *"The cave is dark; our foes are large,*
> *We'll let enlightenment lead the charge!"*

Lady Olwyn flew back toward the cave, pulling Lulu by one hand and Elizabeth by the other.

The giants were not surprised to see them. Cyclops seemed unnaturally sedate as he lay on the bed with his large eye partially closed. The table had been set upright and the colored coins were placed in neat stacks. Lulu noticed that one of the piles held yellow-colored coins, but couldn't see whose face was upon them.

158

Ogias sat on the bed with his elbows on the table and his fingers interlaced. He smiled widely. Lynette was positioned in the crook of his arm.

Lulu stepped forward without hesitation. "I propose another game of skill," she said. "Another chance to win back my friend."

Ogias looked at Grendel and winked. "You lost the first game," he said to Lulu. "But in the name of fairness-" He winked again. "-in the name of fairness, I will give you two more chances. Win both games and your friend will be released."

Lulu eagerly shook her head in agreement.

"But if you lose-" continued the giant. "What will you give to us?"

Grendel sniffed the air and pointed straight at Lulu. "The sword!" he grunted. "She's in possession of a sword I have long sought. The Jabberwock's blood is still fresh upon it." He sniffed the air again, and Ogias followed suit. Both giants' nostrils flared, and Grendel licked his lips with a green tongue. Lulu took another look at Lynette before pulling the Vorpal Blade from her shirt. The Cyclops sat up in bed. A bright glow surrounded his obsidian eye. Lulu reached up over her head and placed the dagger on the table.

The giants grinned at each other. "We have a deal!" boomed Ogias. "Grendel - bring forward The Heart of the Oak." Grendel disappeared to the back of the cavern.

Ogias put his chin in his hands. "A poem while we wait." There was only silence as Lulu and Elizabeth waited for Ogias's poem.

"I asked for a poem!" said the giant.

"Is this part of the challenge?" Lulu asked.

Ogias sat back and laughed. "Is the music of poetry not fit for a giant's ears?" Lady Elfinhart had forgotten to mention that some giants preferred poetry to debates.

Lady Olwyn stepped forward and began reciting in her exquisite voice:

> *"Ipomoea greets the glorious morn,*
> *Unrolling beauty dew has adorned-"*

"What is this nonsense?" roared the giant. "I want poetry I can understand!" He turned his head and spat on the floor. Lady Olwyn's lower lip quivered, and her large eyes took on the look of a frightened doe as she stepped back to join Lulu and Elizabeth. Elizabeth gave her a consoling hug.

Just as Lulu was trying to imagine the type of rhyme the giant would approve of, Grendel appeared out of the darkness. His figure was bent over as he hauled a giant wooden cone on his back. He dropped the heavy cone, set it upright on the floor, and returned to the back of the cave.

Ogias tapped his fingers impatiently on the table. "I'm still waiting for that poem," he boomed.

This time, it was Lulu who stepped forward. As she opened her mouth, she caught sight of Grendel returning with an enormous saw in his good hand. It was the largest saw Lulu had ever seen, and for a moment she tried to recall whether she had ever seen a saw before (as there was little need for one in the suburbs). Lulu composed a poem on the spot:

> *"I've never seen a saw,*
> *A saw I've never seen.*
> *But know I how a beaver looks,*

And how incisors gleam."[50]

Ogias clapped his hands and Lulu had to cover her ears at the noise. "Now *that* is what I call a poem!" Lulu did not dare look back at Elizabeth after borrowing (*or was it butchering?*) her beloved Emily's poem.

Ogias stood up, Lady Lynette still firmly tucked in the crook of his arm, and walked over to the wooden cone. "The Heart of the Oak holds great magic," explained the giant. "Merlin procured this piece himself, and we will cut shields from the wood for the battle ahead." Ogias cleared his throat. "But it is important-"

"Critical," emphasized Grendel.

"Yes, it is *critical* that the face of each shield is a different shape."

"Not the same shape of a different size," the Cyclops elaborated, "but four completely different shapes."

"You mean *three* shields of different shapes," corrected Lulu.

"No - we will need four. We're expecting a-guest." Ogias slapped Grendel on the back, and they exchanged winks.

Elizabeth moved to Lulu's side. "We'll do this challenge as a team," she whispered.

Grendel opened his hand to reveal sticks of charcoal. "I will cut on the lines you draw." Lulu, Elizabeth, and Lady Olwyn each took a piece.

"There will be no second chances," Ogias added. "The Heart of the Oak is irreplaceable. If it is damaged, you will meet an undeniably painful end." He grabbed a stick of charcoal from Grendel's outstretch hand and

---

[50] This original poem is written in the spirit of Emily Dickinson's "*I never saw a moor*"

crushed it in his fist before scattering the fine black powder into the air. Lady Olwyn dropped her charcoal and stepped back. She sat down on the floor of the cave shaking.

"We can do this," said Lulu aloud. She looked over to Elizabeth. She expected to see her sister frightened and shaking as well. Instead, Elizabeth stood tall and confident.

"Conic sections!" she said and immediately drew a line across the top of the cone, parallel to its base. "This shield's face will be a circle."

Next was Lulu's turn. Her line also cut across the cone, but at an angle. "This one will be an ellipse." An unpleasant thought brought a knot to Lulu's stomach. *What if the giants didn't recognize an ellipse and a circle as different shapes?* She decided not to worry her sister with uncertainties.

Elizabeth was already drawing a line from a spot just below Lulu's line diagonally down to the cone's base. "A parabola," she whispered.

Lulu felt her chest tighten further as a lump built up in her throat. She took a deep breath and proceeded to the last line. It cut from a lower spot on the cone down to its base in a line that was parallel to the cone's vertical axis. "A hyperbola." Lulu's voice was barely audible. *How would one explain the difference between a parabola and a hyperbola to a giant?*

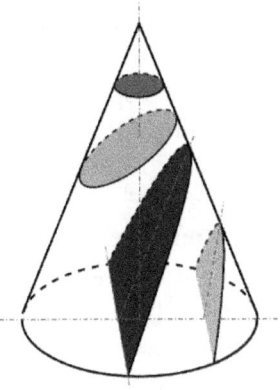

Grendel began working on the shields at once, moving the enormous saw in large even strokes that sent sawdust raining down onto the cavern floor. When he finished, Grendel placed the shields with their faces up - a circle, oval, parabola, and hyperbola.

"The circle's mine!" yelled out the Cyclops, flying from the bed and diving upon the circle shield.

"No, it's mine!" Grendel yelled and began pounding the Cyclops with his wooden arm. Lulu saw something shiny flying across the room, and the Cyclops let out a scream. Ogias sighed and grabbed his two fellow giants - Grendel in one hand and Polyphemus in the other. Lady Lynette dropped from the bed and quickly scrambled across the floor into Lady Olwyn's embrace. Ogias lifted the two lesser giants in the air and bashed their heads together before tossing them both upon the bed. The cyclops's eye was missing. "ENOUGH!" Ogias yelled. He kicked the shields out of the way and turned back to Lulu and Elizabeth.

"It all comes down to the last game," he boomed and rubbed his hands together. "Don't you go anywhere."

Ogias turned and walked to the back of the cavern.

"Let's run for it," whispered Elizabeth. The Ladies were already heading for the door.

"You go ahead," said Lulu. "I gave Ogias my word, and I intend to keep it."

Elizabeth gave her sister a questioning look.

"And I don't want to lose my sword," she admitted and forced a sheepish grin.

Elizabeth looked toward the cave exit and again at her sister. "I'm staying by your side," she whispered.

Ogias returned shortly and didn't seem to notice the Ladies' absence. He placed three identical boxes in front of Lulu and Elizabeth. "What would you say if I told you that one of these boxes holds the Holy Grail?" he asked.

"I would ask to see it before I believe you," Lulu replied skeptically.

"Then you'll have to just take it on faith then, won't you?" chucked the giant. "Choose the correct box, and the treasure of the ages shall be yours. Do you accept the challenge?"

Lulu and Elizabeth exchanged worried glances. "A game of chance," Elizabeth muttered under her breath.

"Can I refuse?" Lulu asked.

Ogias opened his great mouth and roared in response.

"I guess that's a 'no'," said Lulu, trying unsuccessfully to lighten the mood. She squeezed her sister's hand. "Yes, I accept the challenge."

Lulu closed her eyes and took a deep breath. With two hands, she rubbed the parts of her sleeve that had turned gold, hoping that their brief contact with the hazel water might ignite some magical intuitive power. Ogias shifted his weight impatiently. "That one," Lulu finally said pointing at the middle box.

The giant chuckled, but Lulu could not read his expression. He removed the lid from the box on the left and tipped it so Lulu and Elizabeth could see that it was empty. Ogias then moved his finger to the middle box that Lulu had chosen and began tapping it rhythmically. "Would you like to change your selection?" he asked. From somewhere outside the cave, a Great Horned Owl screeched.

Lulu's first instinct was to stick to her initial decision. "One must not change one's core values in changing circumstances," Lady Elfinhart had said. She looked to Elizabeth who quickly whispered: "the odds have changed." *How can the odds change?* wondered Lulu, *there were three boxes, so I had an equal chance - one out of three - of choosing the correct one. Now that there are two boxes, there is a fifty-fifty chance that my box is the right one. Or is there?* Lulu suddenly understood. *If I had a one-out-of-three chance of picking the correct box before, there was a two-out-of-three chance that the prize was in one of the other two boxes. The giant knew which box was correct and that was not the one he revealed. That means that changing my answer would give me a two-out-of-three chance - the odds would improve.* She looked back at Ogias "I change my answer!" Lulu said firmly.

The giant moved his finger to the other box. He lifted the lid and shook the box upside down. It was empty.

Ogias scooped up the remaining box and placed it on the table. "You've lost," he said. "I saw your friends leave, but it is of little consequence. You may go now. We giants have important business to attend to."

Lulu and Elizabeth did not move.

Grendell jumped up. "At last!" He reached for the Vorpal Blade and dropped it at once on the floor with a scream. "I'm burned!" he yelled. "The weapon is cursed!" Lulu reached for the small dagger. It was warm, but not burning. "The blade will reveal its power when needed," Mrs. Magpie had said. *Was this how it would do so?* The owl's screech sounded again.

Ogias shoved Lulu aside, sending her sliding across the cave. Using a cloth, he retrieved the Vorpal Blade and unsheathed it in a single motion. The blade was glowing. Ogias closed his eyes from the glare and brought his face close to the weapon, giving it an audible sniff. Then, holding the blade away from his face, he rubbed it vigorously with the cloth. Just as Ogias the Giant was about to drop the material into the hazel water atop the headboard, Lulu and Elizabeth flew for the exit. They did not want to wait for what was about to happen.

Play Along: In this chapter's "play along" section, you will investigate the riddles posed by the giants: map coloring, conic sections, and the Monty Hall problem. Join the fun on page 258.

14. Battle Plans

A progression of loud screeches met the girls' ears as they exited the cave.

"Run! The Jabberwock is coming!" Elizabeth yelled, her heart beating fiercely. Lulu grabbed her hand as they ran toward a cluster of trees. The screeching grew louder, and Elizabeth added her own frightened shrieks to the noise. "Great - horned - owl," Lulu managed between breaths. "In - the - day." Lulu's words didn't compute in Elizabeth's mind until they had broken through the trees into a clearing. The screeching stopped, and the girls were met with a scene that looked like a familiar painting – '*Washington Crossing the Delaware*' by Emanuel Gottlieb Leutze (1851).

All four Ladies were standing atop the bed with the flowered quilt. Lady Elfinhart was posing stoical, like George Washington, with one foot on the headboard and a sword at her side. Lady Olwyn stood behind her holding the flowered quilt high above her head like a banner. Lady Lynette sat toward the back of the bed gazing far out over the land. Although her lips were pursed together bravely, tears were streaming down her cheeks. Lulu realized how out of place it seemed to compare these Arthurian ladies to an iconic painting of the American Revolution, but this was the image that came to mind. She turned her head to see what was holding the Ladies' interest. Thick black smoke was rising into the air above Hengoen Hill as the white horse, and the Incorruptible Roses looked on.

"Mendel's garden!" Elizabeth yelled out, and the Ladies' trance was immediately broken.

Lady Elfinhart leaped off the bed first, still holding the sword, and ran toward the girls. "We haven't much time," she called out. "The Populus Post is abuzz with activity." The nearby aspens quaked and whispered as if to confirm this. "King Hallden is desperate to find out when and where the Black Pig will launch his attack. The war is yet to start, but dark forces have already breached Camelot!"

The other Ladies were soon at her side. "Merlin's spies are everywhere," said Lady Lynette. "The bedchamber has been ransacked. All the beds were smashed and all the jewels stolen!"

Lady Elfinhart looked again toward the smoke. "We need to see what we can salvage from Menalowan - from Nimue's laboratory. Lady Olwyn, come with me. Lady Lynette, you must find a safe refuge for the girls until the war is over."

"We're coming with you-" said Lulu.

"No." Lady Elfinhart's voice was firm.

"We're ready to fight for Camelot-"

"No," Lady Elfinhart repeated. "We are beyond grateful for all your help." Here her voice softened. "But Camelot is not your home, and this is not your war to fight."

"You mustn't worry about us," Lady Elfinhart continued. "We are not alone, but guided by the knowledge and strength of the multitudes we've met through the beds - Harrison, Linneus, Yeats, Mendel,

Kipling, Chesterton, Dickinson, and so many others."
Lulu noted that there were far more men than women on
that list. *I'm going to change that*, she thought.

Lady Olwyn's beautiful voice filled the air a final
time with Emily Dickinson's words:

> *"Through the Dark Sod - as Education*
> *The Lily passes sure*
> *Feels her white foot - no trepidation*
> *Her faith - no fear*
>
> *Afterward - in the Meadow*
> *Swinging her Beryl Bell*
> *The Mold-life - all forgotten - now*
> *In Ecstasy - and Dell"*[51]

"Wherever lies knowledge, lies hope and
redemption," said Lady Elfinhart.

She motioned to Lady Olwyn, and they set off at
a jog on the narrow path up toward the Hill of Hengoen.
Lulu and Elizabeth watched their departure with mixed
feelings.

Lady Lynette took both girls' hands. "Your bed -
the one on which you arrived - was destroyed along with
the others," she explained. "We have not lost all the
beds, however, nor all the jewels. Some beds were abroad
when the bedchamber was breached. Some gems have
been hidden for safekeeping. It is rumored that Merlin
held a private stash. He dared call it the 'Holy Grail'!"
Lynette scoffed. She didn't hear the girls gasp.

"Queen Nimue will never let the Lake of Tears
fall," she continued. "There is still hope. There will
always be hope." Lady Lynette wiped the tears from her

---

[51] Emily Dickinson, "Through the Dark Sod - as
Education"

cheeks and raised her chin so that she took on the visage of a warrior.

"As Lady Elfinhart said, this is not your war to fight. We must keep you safe for now until an alternate path is found for sending you back where you belong."

Elizabeth sighed. Lulu opened her mouth to protest. She reached instinctively for the Vorpal Blade around her neck before remembering that it was gone.

"The bed in the giant's cave!" Elizabeth suddenly remembered and immediately sprang into action. "Lynette, the other ladies will need your help. You mustn't worry; Lulu and I will find a way home. You know you can trust me." Lynette dove at Elizabeth with a grateful embrace. Elizabeth held her new friend tightly.

Lynette finally pulled away. "Good luck to you both," she said looking earnestly at Elizabeth and then Lulu in turn. "Wherever your adventures lead you, always remember that honor's code is a belief we must preserve until the end." She spun around and began bounding up the path.

"Good luck to you, too," Elizabeth called after her.

"To all of Camelot," shouted Lulu as Lady Lynette disappeared from sight.

"I forgot to tell the ladies to beware of Jabberwocky," said Elizabeth. "I hope they'll be alright." She was shaking.

Lulu put an arm around her sister's shoulder. "You were so brave in front of those three giants. What makes the Jabberwocky any more fearsome than they?"

"The jaws that bite," Elizabeth answered. "Not to mention the claws that catch."

Lulu resisted the urge to roll her eyes.

"And don't forget the eyes of flame," Elizabeth added.

Lulu sighed.

"I wouldn't hold such a fatalistic attitude when it comes to Jabberwocky if I were you," Elizabeth warned. "Up until a couple of minutes ago, you were carrying around the blade that cut off its head. I don't think the Jabberwock would take too kindly to that." A look of sorrow washed over Lulu's face as she remembered that she'd lost her finest treasure.

Just then, a piercing scream filled the air. It was nothing like Lynette's owl call. *The sound of pure evil,* thought Lulu as the girls darted into the cover of the bushes. "Jabberwocky," Elizabeth mouthed. Her heart was racing. The girls sat motionless inside the dense shrubs, barely breathing. They could hear the monster sniff and snort. Branches cracked and split beneath its feet. The aspens quaked. Then the call of a fearless chickadee could be heard from somewhere deep in the forest - "chicka-dee-dee-dee-dee-dee-dee-dee". The monster paused, listening, before howling his response.

Silence. Then more branches splintered in its steps. *What was he searching for?* The chickadee repeated its call, but now it was joined by another voice and soon a third. Lulu imagined an entire army of chickadees approaching the beast, a tiny wood-nymph upon each feathered back. There was no sign of any attack, however, though the birds continued their warning cries from the safety of the trees. "Chicka-dee-dee-dee-dee-dee-dee-dee" Next, a nuthatch added his battle-cry to the song "wha-wha-wha" and within seconds, the entire forest exploded in warning calls.

The Jabberwocky howled. Lulu, with courage from a chickadee, dared to peek out from the cover of the brush. She saw the back of the enormous black beast disappearing up the trail that the Ladies had taken. "Quickly - we need to create a distraction," she said aloud and jumped out into the clearing waving her arms and whistling. The beast - Jabberwocky - was already gone.

A horn blew in the distance, ringing out loud and clear. Lulu dove for cover again as the sound of hoof-beats neared. She peeked out to see a dozen men approaching on horseback. A hoof narrowly missed Lulu's head as Elizabeth pulled her into the thicket. Lulu looked out again to catch a glimpse of the riders heading toward Hengroan Hill. She let out a sigh and lamented that she did not get a good look at the leader of the riders - *was it Hallden? Arthur? Fion?*

Soon all was again quiet. The bird calls had ceased, and only the aspens whispered.

Lulu wiped away the dust which the horses had kicked up into her face. "Let's get out of here," she said. "Should we take the bed back to the castle? Or perhaps the Lake of Tears would be better. Queen Nimue will keep us safe."

Elizabeth looked at her defiantly and shook her head. "We're headed back to the cave."

When the girls re-entered the cave, it was evident that the giants were preparing for battle. Very few humans will ever lay eyes upon a real giant. Only a fraction of those will have the unwelcome opportunity to see a giant in full armor (so you'll have to take Lulu and Elizabeth's word that this is one of the most threatening sights in the world). Ogias, covered in black chainmail, was fitting a helmet onto the Cyclops's head. The helmet had a single slit at its center behind which Polyphemus's

obsidian-eye sparkled. Grendel was sanding his shield -
the ellipse, which he'd already painted green.

"Go away," Ogias boomed when he saw the girls
enter the cave. "We have no more need for you!"

Elizabeth continued walking toward the giants. "I
want a chance to earn back my sister's blade," she called.

"Go away," Ogias repeated, this time not even
lifting his eyes

"A final game," said Elizabeth. "If you win, my
sister and I will be in your servitude."

Ogias looked up.

"On my honor," added Elizabeth holding three
fingers against her heart and hoping that this gesture
meant something to the giant.

Lulu was frightened. "I hope you know what
you're doing," she mumbled under her breath. She took
another look at Elizabeth. Her sister was standing
straight and tall. Lulu hardly recognized the portrait of
the confident young lady standing next to her.

Ogias let his hands drop from the Cyclop's helmet
and took a step toward the girls. His chainmail jingled.

"I have no need for servants, two weak little girls.
But perhaps your lady friends in Camelot will find you of
some value."

Grendel moved toward Ogias and whispered
something in his ear.

"We do have one small matter you can settle."

Elizabeth squeezed her sister's hand.

"I'm ready for my challenge," she declared.

Ogias nodded, his chain-mail jingling, and cleared
his throat. He moved the Vorpal Blade to the edge of the
table and produced a parchment which he held between
his thick fingers. Lifting the parchment close to his face,
the giant read it aloud.

173

"The Black Horse Vale stands 115 kilometers northwest of Camelot Castle," he began.

Lulu pulled out her notebook and started taking notes.

"At precisely eight o'clock tonight, the Black Pig will ride straight toward Camelot at 60 kilometers per hour. He expects to be spotted as he passes Hengoen Hill, which is forty-five kilometers into his trip. Fifteen minutes later, King Hallden will come riding to meet him - at a rate of fifty kilometers an hour. When they meet, the Black Pig will give the signal that the war has begun."

"This is where we come in," said Grendel eagerly. He lifted his shield in the air. Lulu thought she saw a streak of green moving toward the exit of the cave. Indeed, N. Wake was on her most important mission yet.

Ogias cleared his throat and continued. "What we need to know - umm, what you must calculate for your challenge - is how far away from Camelot the Black Pig will cut down the weak king. Where will they meet?"

Lulu handed Elizabeth her open notebook. The page had all the important numbers written in silver glitter. At the bottom, were the words: "I believe in you!"

Elizabeth hesitated. The math was not the problem - it was her conscience that held her back. Elizabeth looked again at Grendel who was admiring his shield. "Haven't we aided the enemy enough?" she whispered to Lulu. "We can't lay out their battle plans, as well."

Lulu, too, was torn. "I'll stand by your side regardless of what happens. The choice is yours." She handed her sister the pen.

Elizabeth fidgeted nervously. Lady Lynette's words echoed in her head: "Honor's Code is a belief we must preserve until the end."

"We need the answer," bellowed the giant.

Elizabeth felt a fire burning in her head, in her belly, in her fingertips. She began to write feverishly. The words seemed to flow out on their own.

Ogias groaned impatiently. "Very complicated calculations," Elizabeth mumbled in response.

Finally, Elizabeth handed the pen back to her sister and faced the giants. "Here is your answer," she said as she handed the notebook to Ogias.

Seeing that the response was written in the form of a poem, the giant smiled broadly and motioned for Grendel and Polyphemus to gather around. Lulu caught a glimpse of green out of the corner of her eye. The sound of clanking armor filled the cavern as the giants moved into position for the recitation. Lulu, meanwhile, edged closer to the table and pulled down the Vorpal Blade. The two girls inched toward the bed. The giants took no notice. Lifting the notebook up to eye-level, Ogias began his performance in a booming voice:

> *The beasts were growling and clawing*
> *beneath an ominous sky.*
> *The men were preparing for battle*
> *as war was drawing nigh.*
> *Something dark was a-stirring*
> *when the castle clocks struck eight;*
> *And the Black Pig came riding-*
> *Riding-riding-*
> *The Black Pig came riding,*
> *headed for Camelot's gate.*
>
> *He rode past the Hill of Hengoen,*
> *the time was a quarter to nine,*
> *In his eyes burned the fires of hatred,*
> *at the sound of the warning bell's chime.*

His lips curled back full of anger,
his unearthly screams raged with spite,
And the Black Pig kept riding-
Riding-riding-
The Black Pig kept riding,
past the White Horse in the night.

The King was awoken from nightmares,
heart and mind full of strife,
He mounted his steed on the hour
and rode for Camelot's life.
The stars were strewn 'bout the heavens
whilst the moon cast a silvery hue,
And King Hallden came a-riding-
Riding-riding-
King Hallden came a-riding,
to fight for all that was true.

The sword of the king was a glimmer,
the Black Pig let out a wail,
The start of the war has been signaled,
will light of darkness prevail?
And where was this crucial encounter
when cold steel fell upon steel-
When the blades of war came clashing
Clashing-clashing-
When the blades of war came clashing?
This, I shall never reveal![52]

Ogias's roar was deafening. Lulu and Elizabeth were already on the bed, but N. Wake was still turning the dials. Grendel threw his shield at the bed, upsetting the

---

[52] This original poem was written in the spirit of *"The Highwayman"* by Alfred Noyes, 1906

cup of hazel water which soaked the green mouse. Elizabeth felt a splash hit her leg. That was the last thing she remembered of Camelot.

Play Along: When and where will the battle begin? Find out in the "play along" section on page 267.

15. Magic

Instead of waking up with a rhyme in her head, Lulu awoke with a compliment on her tongue. "Your poem was great," she said, turning toward Elizabeth. Her sister was still fast asleep.

"You're back!" came a familiar voice.

Elizabeth turned her head to see Petrus standing next to the bed. He wore a grin that stretched from one end of his round face all the way to the other. In his hand, Petrus was holding a fishbowl. Its occupant was a Siamese fighting fish (a betta) who flared her long iridescent tail and dove at the glass bowl again and again. To Lulu, the fish bore an uncanny resemblance to someone.

"My friend the Kraken delivered her this morning," explained Petrus, tapping a fat finger against the fishbowl. "And it isn't even a Thursday! I call her Queenie, or Q for short."

Lulu returned the bald man's smile.

"Kraken also brought a message from a dear, dear friend," continued Petrus. "I've been promoted to Captain," he beamed. "You're looking at the new Petrus – a pilot of progress, an authority on action, an executive of exacting retribution (or a vigilante of vengeance, if you will). The Outpost does not look kindly on those who destroy its property, including certain beds."

Lulu waited for a break in the soliloquy. "Congratulations," she offered.

"But look at us, chit-chatting away. As the new Captain, I must enforce the rules. You and I, we are no lingerers, you know."

Lulu felt the familiar drops hit her eyes. *I always expected heroism to appear in the form of a brave knight,* Lulu thought before the Outpost faded from view. *But now, I will never discount the true fortitude of three Lily Maidens, the gallantry of a tiny green mouse, nor the mettle of a short, bald man at Fifty North and Forty West. And that's magic…*

16. New Buds

"Was it all a dream?" asked Lulu. "A strange dream?" She had awoken not in bed #84, but beneath the starched white sheets of her own little bed at the Lake Despinassy cabin - a bed with no gems. Sunlight was pouring in from the east-facing window. Scratching sounds were coming from the other side of the room. Lulu sat up. The shag rug had been thrown aside, and Elizabeth was frantically clawing at the floor. The cellar door had vanished.

"It couldn't have been a dream," Elizabeth cried.

"That's the second worst way for a story to end - a sign of a lazy author," agreed Lulu. "Except for 'Alice in Wonderland', of course. Lewis Carroll was a genius." She instinctively reached for her notebook, first searching her pockets and then groping around the nightstand.

Elizabeth watched her sister fumbling for a moment before she remembered something. "I left your notebook with the giants," she said.

"So it was real after all!"

"There's no empirical evidence-" Elizabeth started. "But I have - a hunch-"

The twins stared at one another, each comparing her story with the one projected in her sister's eyes.

"Do you think the battle was successful?"

"What if Camelot fell and the last of the beds and jewels were destroyed? Since the events happened in the past, maybe that is why the cellar and the ship-bed are no longer here. It's like they never existed." Elizabeth's voice faded with those last words.

"Or maybe it has something to do with the Fairy Queen's wish," suggested Lulu.

"I guess we'll never know," said Elizabeth quietly.

Lulu thought of the look on King Hallden's face and shook her head. "I have a feeling that there was a happy ending to the story. Good always triumphs over evil."

"Who says the story is over?" came a voice.

Lulu and Elizabeth turned to the right. A small table that had not been there before had been pushed up against the wall. The girls ran to the table and found a vase that held three lilies. "Is there a note?" Elizabeth asked. Lulu pulled the vase toward her. Behind the vase was a small gray Mouse. Elizabeth did not even flinch.

"N. Wake?" Lulu asked.

"Call me Nancy."

"But how did you-"

"I made a wish -" said the Mouse, anticipating their questions. "When the hazel water spilled, I was touching a gem."

Elizabeth suddenly remembered the water that had spilled on her leg. Her pant leg had turned gold, and when she pulled it up, she saw a coffee-colored mark on her calf. Later, when others would comment on this 'birthmark', Elizabeth would just nod and smile.

"Nancy is right," said Elizabeth, patting the Mouse on her head (the Mouse would have normally taken offense to being treated like a mouse, but on this occasion, she made an exception - for Elizabeth). "I have a feeling that Mrs. Magpie has more adventures in store for us."

Lulu's eyes shone. "Mrs. Magpie!" She reached into her pocket and pulled out Mrs. Magpie's note.

"If you can fill the unforgiving minute
With 60 seconds' worth of distance run,
Then yours is <u>Wonderland</u> and everything that's in it
And which is more, you'll <u>earn your sword and crown</u>."[53]

"I feel like we've grown up so much in our brief time in Camelot," Lulu sighed. "Do you think that counts as 'filling the unforgiving minute with sixty seconds' worth of distance run'?"

Elizabeth shrugged.

"If we did win the race, I'd take the sword," said Lulu, feeling for the Vorpal Blade around her neck. "But I'm not sure I want the crown anymore," she added, recalling the weight of the Fairy Queen's crown.

Elizabeth reflected on their time in Camelot. She couldn't quite put words to it, but somehow Camelot felt like an in-between place - a transition, a training ground. *Does a butterfly dream within her chrysalis?* she wondered. Elizabeth shrugged off these thoughts and shook her head. "Actually, I don't think the race is over- not yet. I believe that we're only about two seconds in." She looked down at Nancy who was now asleep in her hand.

Lulu read Mrs. Magpie's note once again, silently this time. A piece of lint from her pocket was clinging to the paper, just below the last line of the poem. Lulu brushed it away with her sleeve. She stepped back in astonishment. A drawing of a trefoil knot appearing

---

[53] Rudyard Kipling, "*If—*", 1895. Underlined changes by Mrs. Magpie.

before her eyes. "The hazel water!" gasped Lulu as she inspected the golden splatters on her sleeve. She rubbed it firmly against the paper. Words appeared.

"Look!"

Elizabeth was soon looking over Lulu's shoulder as the new clue was revealed.

> *"Cities and Thrones and Powers*
> *Stand in Time's eye,*
> *Almost as long as flowers,*
> *Which daily die:"*[54]

"That's a sad poem," said Elizabeth with a sniffle, thinking of the Lily Maidens.

Lulu continued rubbing. "There's more!"

uhmendig, wretdree, nfdoadTi, usonpoaii, aulunsts, btwhueCa, stattihr, BdTescEta, eonfnehs, tpgOenren

Below the letters, there was a drawing in the shape of a key. Within it, were the words "the Earth."

"I need to write down this clue for my readers. Do you think they'll solve my puzzles along with the book?" Lulu reached for her notebook and then remembered that it was gone.

"I'm sorry about your notebook," said Elizabeth.

"Not a problem," said Lulu, unzipping a backpack. It was filled with notebooks and pads in an assortment of sizes and colors.

"But the story you were working on-"

Lulu smiled as she pocketed a small pink notepad. "Math and Magic in Camelot was always Mom's story. I

---

[54] Rudyard Kipling, "*Cities and Thrones and Powers*", 1906.

need to find my own voice as an author. William Butler Yeats said it best:"

> *"I made my song a coat*
> *Covered with embroideries*
> *Out of old mythologies*
> *From heel to throat;*
> *But the fools caught it,*
> *Wore it in the world's eyes*
> *As though they'd wrought it.*
> *Song, let them take it,*
> *For there's more enterprise*
> *In walking naked."*[55]

Elizabeth raised an eyebrow.

"In case you were wondering, the only thing I plan on baring is my soul," added Lulu and they both laughed and laughed. It wasn't so much at the hilarity of the joke, but the laughter brought with it a sense of relief that they could still be silly and fun, that they didn't quite have to grow up too quickly. They were still allowed to do as much roving as their hearts desired.

Elizabeth was the first to sober up from the laughter.

"Fans of Mom's first book wouldn't have cared for the sequel, anyway," she said. "It's too dark, too grown up."

"Not enough puns," Lulu smirk. "I should have made some witty jokes as the Fairy Queen was trying to kill me."

"Too much poetry and not enough math," added Elizabeth "And profiles of real people alongside myths. It's all very confusing."

---

[55] William Butler Yeats, "A Coat", 1913

"But my brilliant readers-"

"You can't win fans with flattery," Elizabeth interrupted.

Lulu considered this for a moment. "Like I said, this is Mom's problem, not mine. Let's get back to the clue."

Elizabeth examined the paper. "It appears to be a cipher."

At this, the Mouse perked up her ears. "Did someone say 'cipher'?" she asked. "I'll have you know that, green or not, I am still the greatest Mouse spy that has ever set foot in- where exactly are we?"

"Lake Despinassy," replied Lulu.

"In Quebec, Canada," Elizabeth clarified. "We're on vacation."

Nancy jumped onto Lulu's shoulder and looked at the note. "I believe it might be a poetry code," she said. "But we need the key."

"The words 'the Earth' are written inside the drawing of a key. Could that be it?"

The Mouse shook her tiny head. "That's too simple. Spies never make codes that are so easy to decrypt. The key would be a word (or multiple words) from the poem."

"I know!" Elizabeth yelled out. "The first part of the note is a poem called "If" by Rudyard Kipling. Mrs. Magpie changed some of the words."

Lulu began jumping up and down excitedly (as Nancy held on for her life). "Instead of *yours is the Earth*, Mrs. Magpie wrote *yours is Wonderland*. Wonderland is the key! I just know it!"

The Mouse looked to Lulu and then back to Elizabeth. Elizabeth smiled. "Let's give it a try," she said.

Lulu wrote down the word "Wonderland."

185

"The first step is to assign a number to each letter in the order they appear in the alphabet," explained the Mouse. "If two instances of the same letter appear, the first one from the left gets the lower number, and the next one gets the next number."

Lulu cocked her head to the side. "So 'A' is number 1?" she asked.

"Yes," confirmed Nancy. "Since 'B' and 'C' don't appear, the first 'D' is number 2, and the second 'D' is number '3'. Keep going until you've assigned a number from 1 to 10 to the ten letters in 'WONDERLAND'."

W	O	N	D	E	R	L	A	N	D
			2				1		3

"I get it!" beamed Lulu as she began working on this task.

"What's next?" Elizabeth asked as soon as her sister had finished writing 10 beneath the "W".

"Write the numbers from 1 to 10 across your paper. Each group of letters in the encrypted message is a column. The numbers you wrote beneath WONDERLAND will tell you which column goes with which letter group."

Lulu looked down at her paper. "I assigned number 10 to the letter W, so the first group is column 10?"

"Correct."

uhmendig, wretdree, nfdoadTi, usonpoaii, aulunsts, btwhueCa, stattihr, BdTescEta, eonfnehs, tpgOenren

1	2	3	4	5	6	7	8	9	10
									u
									h
									m
									e
									n
									d
									i
									g

Deciphering the code was easy and fun.[56] When Lulu was done, her sister read the rest of the poem aloud and smiled.

Lulu smiled back. "That was the right sort of poem, if you know what I mean."

"I believe I do," said Elizabeth. "I believe I do."

"So… the story isn't really over. We do have more adventures in store!"

"It's our turn to defend honor's code," said Elizabeth. Suddenly remembering something, she reached into her pocket. "And it looks like Camelot left us with a little Amethyst remembrance![57]" Elizabeth opened her hand to reveal a jewel – one of the two amethysts she'd grown in Mendel's garden. It was still glowing brightly with the energy of youth.

---

[56] Go to page 270 to learn about poem codes and decrypt Mrs. Magpie's message.
[57] She is referring to Emily Dickinson's poem "I held a Jewel in my fingers—"

Lulu danced around the room. "We have so many preparations to make! What are you waiting for?"

Elizabeth moved to the vase of flowers. "The Lily Maidens' behavior was a little obscure at times, but their hearts were in the right place. If the Ladies taught me anything, it was that we don't have to always rush. It's alright to take the time to smell the flowers." She leaned down and let the fragrant aroma reach her nose. "Sometimes it's even appropriate just to pause and recite a poem."

"Or tell a story," added Lulu.

"Or scour your horse." Elizabeth smiled at the quizzical expression on her sister's face. "I'll tell you about it later."

Lulu put a foot atop her bed and posed dramatically.

"We are the Twin-flowers of the House of Lovelace and the Sisterhood of… glitter pens!"

Elizabeth put two fingers over her heart. "Defenders of truth, honor, and the scientific method!"

"Can our outfits be pink?" Lulu asked. And now it was Elizabeth's turn to roll her eyes and laugh.

The end.

If—

By Rudyard Kipling

If you can keep your head when all about you
Are losing theirs and blaming it on you,
If you can trust yourself when all men doubt you,
But make allowance for their doubting too;
If you can wait and not be tired by waiting,
Or being lied about, don't deal in lies,
Or being hated, don't give way to hating,
And yet don't look too good, nor talk too wise:

If you can dream - and not make dreams your master;
If you can think - and not make thoughts your aim;
If you can meet with Triumph and Disaster
And treat those two impostors just the same;
If you can bear to hear the truth you've spoken
Twisted by knaves to make a trap for fools,
Or watch the things you gave your life to, broken,
And stoop and build 'em up with worn-out tools:

If you can make one heap of all your winnings
And risk it on one turn of pitch-and-toss,
And lose, and start again at your beginnings
And never breathe a word about your loss;
If you can force your heart and nerve and sinew
To serve your turn long after they are gone,
And so hold on when there is nothing in you
Except the Will which says to them: 'Hold on!'

If you can talk with crowds and keep your virtue,
Or walk with Kings - nor lose the common touch,
If neither foes nor loving friends can hurt you,
If all men count with you, but none too much;
If you can fill the unforgiving minute
With sixty seconds' worth of distance run,
Yours is the Earth and everything that's in it,
And - which is more - you'll be a Man, my son!

About the Author

Lilac Mohr is a senior software engineer with a passion for problem solving which she hopes to pass on to her five children. She is the editor of the poetry anthology "*Classic Poetry for Your Little Genius*," and the author of the "*Math and Magic*" series of math adventure novels. Lilac holds a B.S. degree in Computer Information Systems and an M.S. degree in Statistics.

> *If you enjoyed this book I wrote,*
> *(If it succeeded in floating your boat),*
> *Please leave a review-*
> *It's the least you can do*
> *To cast your Amazon "vote."*

Play Along

Chapter 1 (answers on page 273)

Topic 1.1: Homing Pigeons

In this chapter, Mrs. Magpie's note was delivered to Lulu and Elizabeth by messenger pigeon. Homing pigeons have a natural ability to find their way home across long distances, and humans have been using them for both sport (pigeon racing) and message delivery for thousands of years. Although some homing pigeons have been trained to fly between two locations successfully (one which is their home and the other offering food), messenger pigeons do not typically perform round-trip deliveries like the Pigeon in the story. He is a Pigeon (capitalized) rather than a pigeon because of the anthropomorphic (human-like) traits he exhibits.

The Pigeon delivered his message at 10 am on a Sunday, and still has a 2,700-kilometer journey ahead of him. If he can maintain a constant speed of 75 kilometers per hour (with no breaks), when will he arrive home to Wonderland (to gossip about the "mercurial squabs")?

Topic 1.2: Cher Ami

One of the most famous messenger pigeons in history was "Cher Ami", a homing pigeon used by the Allied forces during the First World War.

In October of 1918, the 77th Infantry Division, led by Major Whittlesey, became isolated from the other forces and were surrounded by German enemies. Desperate for help as bullets were flying at his men,

Major Whittlesey sent a messenger pigeon to deliver word of their current status.

The pigeon was shot and killed before his message could be delivered. A second messenger pigeon met the same fate. Finally, Major Whittlesey had only a single pigeon left - Cher Ami. She was their last hope!

As Cher Ami flew, carrying with her, not only a message but also the fate of 200 men, she, too, was shot down. Miraculously, the pigeon took to the air again and continued her flight with Major Whittlesey's message dangling from a shattered leg. Cher Ami survived and was fitted with a wooden leg to replace the one that had been injured. The heroic bird was awarded the Croix de Guerre Medal for her wartime service. Cher Ami's body is currently on display at the Smithsonian.

By assigning Elizabeth the moniker of "Cher Ami", the messenger pigeon who visited Lake Despinassy was paying her quite a compliment. As the story continues, do you think Elizabeth will display the perseverance and dedication of her namesake?

Look for this book at your local library to learn more about Cher Ami's story:

Cher Ami: WWI Homing Pigeon
By Joeming W Dunn
Illustrated by Ben Dunn
2012

Topic 1.3: Magnetoreception

Scientists believe that rock pigeons' innate homing ability relies on magnetoreception, a sense which allows them to use the Earth's magnetic field to navigate. Other organisms including bacteria, worms, hens, sharks, and bats have also exhibited magnetoreception to orient themselves. Although humans don't possess this special sense, they can also use the Earth's magnetic field for navigation – with the aid of a compass!

It's very easy to make your own compass at home:

1. Gather your supplies. You will need a sewing needle, a magnet, a piece of wax paper, and a bowl of water.

2. Magnetize your sewing needle by rubbing it repeatedly with the magnet in a single direction (from the end of the needle to the tip).

3. Thread your magnetized needle through a piece of wax paper and place it to float in the bowl of water. Voila! The tip of the needle will always point to the north.

Chapter 2 (answers on page 273)

Topic 2.1. Patterns in Poetry: Rhyme Scheme

"Where is the math?" you wonder. "I thought I was reading a math book, not a poetry book!" Have patience, Dear Reader, and I will let you in on a secret… Math is everywhere!

Mathematicians are always looking for patterns in the world. If you, too, begin to think like a mathematician, you'll see that patterns are hidden everywhere- even in unexpected places like poetry.

The easiest pattern to locate in a poem is the rhyme scheme. Look for the rhyming words in this excerpt from the poem "*How Doth the Little Crocodile*" by Lewis Carroll (1832-1898):

How doth the little <u>crocodile</u>   ←A
Improve his shining <u>tail</u>   ←B
And pour the waters of the <u>Nile</u>   ←A
On every golden <u>scale</u>!   ←B

The first and third lines rhyme, as do the second and fourth lines. This is called an alternating rhyme scheme (or ABAB).

To identify the rhyme scheme of a poem, you may find it helpful to copy the poem onto a separate piece of paper and use different colored markers to underline the words which rhyme.

Match each poem with its rhyme scheme:

1. AABCCB   2. ABAAB   3. AABBA

From *A Book of Nonsense* by Edward Lear

There was an Old Man with a beard,
Who said, 'It is just as I feared!
Two Owls and a Hen,
Four Larks and a Wren,
Have all built their nests in my beard!'

From "*The Road Not Taken*" by Robert Frost

Two roads diverged in a yellow wood,
And sorry I could not travel both
And be one traveler, long I stood
And looked down one as far as I could
To where it bent in the undergrowth;

Topic 2.2. Knot Theory – The Trefoil Knot

The girls recognized the symbol on the headboard (and the charms given to them by Mrs. Magpie) as a trefoil knot. The trefoil knot is the simplest example of a mathematical knot. Unlike the knot you use to tie your shoes, mathematical knots (known as "nontrivial" knots) cannot be untied.

Part A

Take a rubber band (or connect the ends of a piece of string to form a closed loop), and twist it so it looks like this diagram:

Can you "untie" (untwist) the shape so that it returns to its original form? The closed loop is called an "unknot" because regardless of how you twist, bend, or tie it, it can always be untied. Give it a try!

Now make a trefoil knot. First, take a piece of string and tie a regular overhand knot like this:

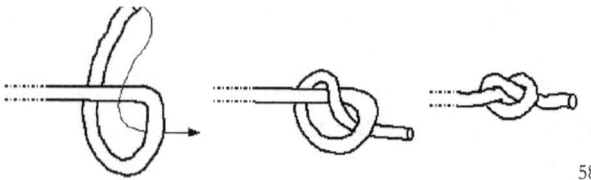

Next, connect the two ends together with a piece of strong tape (like duct tape). You've created a trefoil knot:

Adjust your string so you can see its three intersections. The trefoil knot is named after the three-leaf clover (called a trefoil plant) because it has a crossing number of three.

Try to untie your trefoil knot, so it turns into a closed loop like the unknot (without cutting the string). It's impossible!

Part B

In the book "*Math and Magic in Wonderland*", Lulu and Elizabeth made Mobius strip bracelets by giving their tickets a twist before connecting the ends together. You can create a trefoil knot from a Mobius strip as follows:

1. You'll need a long strip of paper. You can make one from ordinary printing paper.

2. Hold one end of the strip securely (or tape it to the table) while you give the other end <u>three half-turns</u> (note that this is different from a first-order Mobius strip where you give the paper only one half-turn).

3. Tape the ends of the strip together.

4. Using a pen or marker, draw a line that goes through the center of your Mobius strip lengthwise.

5. Cut along this line to create a trefoil knot (no additional tape will be required)!

6. Repeat the process using a Mobius strip created with five half-twists. What do you get?

Topic 2.3. Nancy Wake

N. Wake, the green spy-Mouse introduced in this chapter, was named after Nancy Wake, a decorated World War II spy. If you're interested in reading more about Nancy Wake, look for this book in your local library.

> _The White Mouse: The Story of Nancy Wake_
> By Peter Gouldthorpe
> 2015

The illustrated book tells Nancy Wake's story in an engaging manner which both older children and adults will enjoy.

Topic 2.4. Bio-fluorescent Mice

What does a certain Mouse have in common with a jellyfish? As N. Wake joked, they are both "brilliant"! There are several organisms that are naturally bioluminescent, which means they use a chemical reaction in their bodies to glow (which is also how a glow stick works!). Fireflies and glowing jellyfish are two examples of bioluminescent creatures. But what about mice?

Scientists have used the gene from a bioluminescent jellyfish to develop a technique that will make a mouse glow. The gene, part of the jellyfish's DNA "recipe" contains instructions for green fluorescent protein. Unlike bioluminescent organisms that produce their own light, however, the resulting green mice are fluorescent. This means that they absorb light from an external source and then release it gradually (the way glow-in-the-dark stickers work!).

Why would scientists create fluorescent green mice? The green fluorescent protein (GFP) can be used like a highlighter that lets scientists track traits in the mice. The fluorescent color can help scientists track the spread of cancer inside the mouse's body, for example. This helps them better understand the disease and evaluate the efficacy of different treatments.

Explore bioluminescence and biofluorescence in the natural world on this website:

LuminescentLABS
http://luminescentlabs.org/science.html

You can own a biofluorescent pet! GloFish® are zebrafish genetically modified with fluorescent proteins. They were initially developed to detect water pollution, and are now sold as pets!

The reason for N. Wake's green glow is not immediately clear in the story. Maybe she underwent genetic engineering to make her an extraordinarily intelligent spy-Mouse, and the fluorescence tracks the expression of these new traits.

Topic 2.5. Triboelectric Effect

When Elizabeth rubbed the symbol on the headboard with her finger, she wondered if it was glowing because of a triboelectric effect. The triboelectric effect is when you create an electrical charge by rubbing two materials together. You've most likely experienced triboelectric charging before – it's static electricity!

Take items made of different materials into a dark room and see what happens when you rub them against each other. Some items to try: a comb, a blanket, a balloon.

Topic 2.6. Copyright Law, Parody Poems, and *The Bed Book*

Part A

All the famous poems quoted in this book are part of the "public domain", which means that they are not subject to copyright protection. They can be used or performed freely by the public (with proper accreditation to the original author, of course). All works published before 1923 are in the public domain (what a treasure trove!). Search these public domain repositories to discover some beautiful children's literature (completely free!):

The Baldwin Online Children's Literature Project
www.mainlesson.com

Unfortunately for Lulu, <u>*The Bed Book*</u> was published in 1976 and is still under copyright protection. In the United States, works published before 1978 have a maximum copyright duration of 95 years from the publication date. In which year will Lulu be allowed to quote The Bed Book in her novel?

<u>Part B</u>

Are you among the "smart, curious, resourceful readers" which Lulu described? If so, you may have already located <u>*The Bed Book*</u> to see what makes it "the right type of poem" for this novel. Look for it in your local library:

> <u>The Bed Book</u>
> By Sylvia Plath
> Illustrated by Emily A. McCully
> 1976

You will immediately notice that the poem Lulu has composed (which she recites at the end of the second chapter) uses the same rhyme scheme and meter as <u>*The Bed Book*</u>. This is an example of a parody poem- a poem which imitates another poet's style in a humorous manner. Lewis Carroll, the author of <u>*Alice in Wonderland*</u>, was a master of the genre. Many of his parody poems gained greater fame than the original works they mimicked. For example, compare the first stanza of Carroll's poem "*How Doth the Little Crocodile*" (in Topic 2.1) with the first stanza of the poem "*Against Idleness and Mischief*" by Isaac Watts:

How doth the little busy bee
Improve each shining hour,
And gather honey all the day
From every opening flower!

A wonderful way to uncover even more patterns in poetry is to write your own parody poem. Choose a poem, nursery rhyme, or song to imitate in your own work.

Part C
In *The Bed Book*, Sylvia Plath describes different types of beds – a snack-dispensing bed, an elephant bed, a bed that fits in your pocket, a tank bed, and many others. For a creative writing exercise, describe your own unique type of bed. Draw a picture or create a model of your original bed idea.

Chapter 3

Topic 3.1: Geographic Coordinates

A geographic coordinate system lets you find any place on Earth using a pair of numbers – longitude and latitude. Latitude identifies your position north or south of the Equator. Longitude is your position west or east of the Prime Meridian which runs through Greenwich, England.

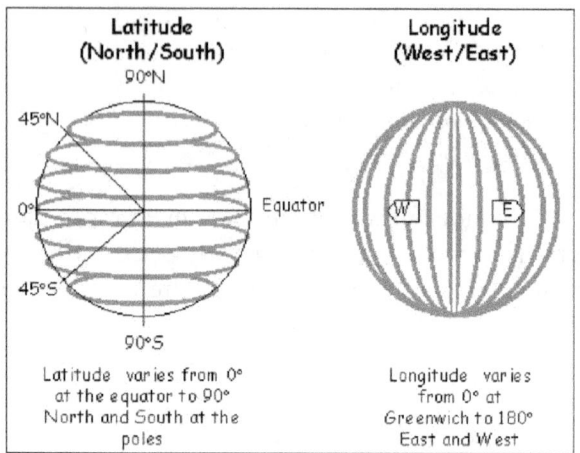

Latitude (North/South)	Longitude (West/East)
90°N 45°N 0° — Equator 45°S 90°S	W — E
Latitude varies from 0° at the equator to 90° North and South at the poles	Longitude varies from 0° at Greenwich to 180° East and West

Now, the "W" and "N" symbols on the dragons' necks should start making sense. Lulu and Elizabeth had to use the dials to enter the coordinates to "Magic". The dials were already set to the bed's current coordinates at Lake Despinassy in Quebec, Canada (49° North, 77.47° West).

Find a map of North America with labeled longitude and latitude lines. Use the map to locate the girls' approximate position.

Topic 3.2: John Harrison and the Longitude Problem

To know your position before a collision
You need the precision of Harrison's vision.

This chapter gave a brief overview of how John Harrison dedicated his life to the development of a clock that could maintain its accuracy at sea. This solution to the longitude problem finally earned Harrison the financial prize from the British parliament in 1773 at the age of 80.

To read more about John Harrison and the longitude problem, look for these books in your local library:

The Discovery of Longitude
By Joan M Galat
Illustrated by Wes Lowe
2012

Sea Clocks: The Story of Longitude
By Louise Borden
Illustrated by Erik Blegvad
2004

Topic 3.3: Using a Clock to Find Longitude

How does an accurate clock help a sailor calculate his longitude? The calculation is surprisingly simple:

Part A

Longitude lines run from the North Pole to the South Pole. The longitude 0 line, called the prime meridian, runs through Greenwich, England. Since our planet is (almost) a sphere, it is convenient that there are 360 lines of longitude, like the three hundred and sixty degrees of a circle (cut an orange in half horizontally, and you'll see that the cross-section is a circle).

Use a globe (or any spherical object like an orange or a ball) to represent Earth. A flashlight can be the sun. As you shine the flashlight on your Earth, note which sections of the planet are in the light (daytime) and which are in the dark (nighttime). Earth takes 24 hours to rotate around its axis (1 day), so every hour, the sun moves 15 degrees across the sky (360 degrees longitude divided by 24 hours in a day).

Let's say that before leaving port, a ship's captain sets his trusty clock (built by John Harrison, of course) to 12:00 when the sun is highest in the sky (solar noon). On the next day, the captain notes that when the sun is highest in the sky, his clock reads 10:30 a.m. (an hour and a half earlier than the previous day). All he has to do is multiply 1.5 hours by 15 degrees longitude-per-hour to see that he has traveled 22.5 degrees longitude to the east.

Try this:

1. Find the current time in the Royal Observatory in Greenwich, England by typing "Greenwich Mean Time" in an internet browser.

2. Find the difference between the time on your clock and Greenwich Mean Time (you may need to adjust for Daylight Savings Time).

3. Multiply this number by 15 to get your approximate longitude.

Part B

The longitude you calculated in Part A is only an approximation of your position. As Lulu explained, time zones assign the same time to an entire geographic region, so the time on your watch (based on time-zone) and solar time (based on the sun) are not always the same. Here is Lulu's example to illustrate the concept:

Locate the U.S. cities of Denver and Salt Lake City on a map. Denver and Salt Lake City are both in the same time-zone (Mountain Standard Time), so people in Denver and people in Salt Lake City will have their watches set to the same time. SLC is located about 7 degrees of longitude west of Denver, however. The sun will reach its highest point (solar noon) to an observer in Denver about half an hour (27 minutes and 36 seconds, to be precise) before it reaches its highest point to a Salt Lake City observer.

For a more accurate calculation of your longitude, you need to find solar noon in your location:

1. On a sunny day, place a stick in the ground and track its shadow to make your own sundial.

2. Starting at about an hour before noon (adjust for daylight savings), record the shadow's length at regular intervals (every 5 minutes). Write down the time where the shadow was at its shortest. This is your solar noon.

3. Set a clock to "apparent solar time" based on the solar noon you identified with your sundial. Find the difference between solar time and Greenwich Mean Time (which you can look up using an internet browser).

4. Multiply the difference by 15 to get a new estimate of your longitude.

5. Find your location on *maps.google.com* and right-click to see your actual coordinates. How accurate was your longitude calculation?

Topic 3.4: Finding Your Latitude with Polaris

Part A

In the last exercise, you calculated your longitude using solar time. Longitude is only one part of your geological "address", however. To identify your location on Earth, you also need to find your latitude. Latitude lines circle the globe horizontally. Your longitude tells you how far north or south of the equator you are.

> "*I must go down to the seas again,*
> *to the lonely sea and the sky,*
> *And all I ask is a tall ship*
> *and a star to steer her by.*"[59]

Sailors have long used Polaris (the North Star) to find their latitude in the northern hemisphere. The North Star stays in a fixed position in the night sky – directly above the North Pole. On a clear night, latitude can easily be calculated by finding the angle between the North Star and the horizon.

To demonstrate this concept, take a globe (or another spherical object like a large beach ball) and ask a friend to hold a small object (like a marble, grape, or plastic toy star) above the North Pole to represent Polaris. Look through a drinking straw at Polaris from different locations on the globe and note how the angle of the straw changes.

At the North Pole, Polaris will be straight above you at a 90° angle from the horizon (latitude 90° N). At the equator, Polaris will be lined up with the horizon (latitude 0° N). In the southern hemisphere, you will not be able to see the North Star at all (so you will not be able to find your latitude using Polaris).

Part B

Now that you understand how to find your latitude using the North Star, let's try it out. The

---

[59] John Masefield, "*Sea Fever*"

astrolabe is an instrument that was invented by the ancient Greeks around 200 BC for exactly this purpose.

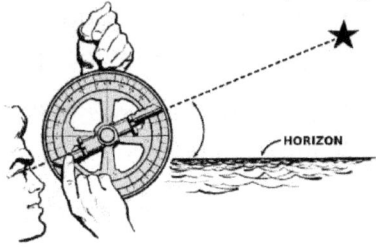

You can make your own astrolabe at home using a protractor, drinking straw, string, and weighted beads:

1. Thread the string through the hole in the center of the flat edge of the protractor.

2. Secure a weight (such as a heavy bead) to the other end of the string.

3. Tape the drinking straw to the flat edge of your protractor.

4. Locate Polaris: Fing the Big Dipper and then following the line of the two stars at the end of the Big Dipper's "scoop" to the bright North Star.

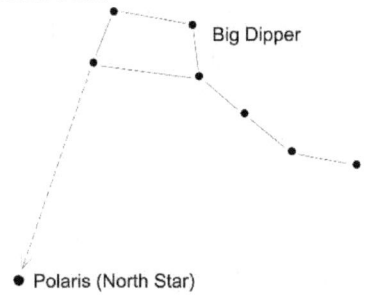

5. Look through the drinking straw on your astrolabe until you can see Polaris. Record the number on the protractor that the weighted string is intercepting.

6. Subtract this angle from 90° to get your latitude.

7. Use a map, globe, or internet website to find your true latitude. How accurate was the measurement you took with your homemade astrolabe?

Chapter 4 (answers on page 273)

Topic 4.1: Half-Way There

Lulu and Elizabeth began their trip at coordinates 49 North, 77.47 West. Fortunately, the twins will not stay in the middle of the ocean for long, since Fifty North and Forty West is only the half-way point of their journey (*"fear not a life pelagic as the distance will double"*). After completing the second half of the trip (moving an equal distance in the same direction), at what coordinates will they end up?

Find these coordinates on a map. What modern-day landmark is in the area?

Topic 4.2: Nomograms

Part A

A nomogram (sometimes called a nomograph) is a graph that allows you to make calculations. Petrus, the self-proclaimed Constable of Calculations, might use nomograms for computing figures related to ocean currents and waves. Consider the following nomogram (Sverdrup-Munk-Bretschneider Nomogram) from the National Oceanographic and Atmospheric Administration (NOAA):

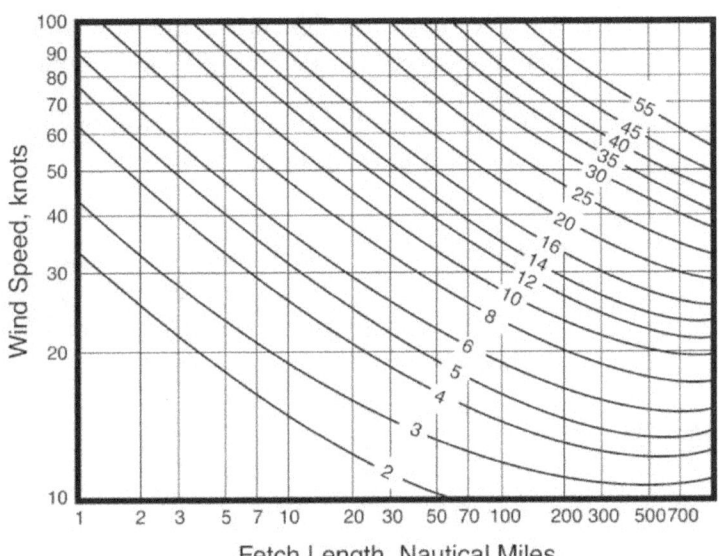

This nomogram allows you to predict the height of ocean waves (in feet) based on wind conditions (assuming the wind blows long enough for the waves to develop). The vertical (Y) axis is the wind speed, while the horizontal (X) axis is the fetch length or the distance that the wind is blowing across the water (at a consistent speed and direction).

If the wind speed is 30 knots and the fetch length is 20 nautical miles, what is the predicted wave height? To find the answer using the nomogram, locate 30 on the Y-axis, and 20 on the X-axis. Follow the lines to their intersection, and you'll see that it is on the curve labeled

"6". Under those conditions, the ocean waves would be approximately six feet high. Use a tape measure to see how high a six-foot wave would be.

Answer the following questions using the nomogram:

1. What is the wave height when the wind speed is 40 knots, and the fletch is 30 nautical miles?

2. If the fletch is 300 nautical miles, what wind speed is needed to produce waves that are 16 feet high?

3. If the wind speed is 70 knots, what fletch would produce waves that are 20 feet high? Use a tape measure to see exactly how high a 20-foot wave would reach on your wall.

Part B

In the story, Petrus's "invention" is a multiplication nomogram that looks like this:

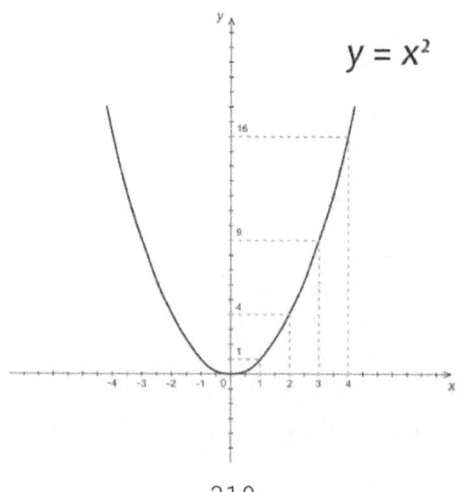

The curve is a parabola: $Y=X^2$. Plot this curve using graph paper. If you read *Math and Magic in Wonderland*, you may remember that square numbers form squares if you use blocks (or crackers). X^2 represents the number of blocks in a square that has X rows and X columns ("X times X"). To plot $Y=X^2$, make a chart of integers and their squares (you may use a calculator):

X	$Y=X^2$
-2	4
-1	1
0	0
1	1
2	4

Your own chart can go from X = -10 to X =10. Plot each point on your graph paper. For example, find -2 on the horizontal axis (X) and 4 on the vertical axis (Y), then place a dot at (-2, 4). After you've plotted some integers, connect the points to form a smooth curve.

Test out your multiplication nomogram by picking two integers from 0 to 10 (6 and 3, for example). Find the first integer on the X-axis and move your finger up to find the point on the curve - in the example, this is (6, 36). Add a negative sign to your second integer and repeat – in the example, you'll find the point (-3, 9). Connect the two points with a ruler. Where does your line intersect the Y-axis? In the example, this point is at (0, 18), so 6 x 3 = 18. The multiplication nomogram worked! Choose different pairs of integers and test it out.

Is this the most efficient way to multiply two integers? Of course, it isn't! Mathematicians love to find patterns and play with numbers. While nomograms were used for many years (before the invention of the computer) for performing navigational calculations, this multiplication nomogram is just for recreation (fun).

For an interesting mathematical craft, recreate Petrus's multiplication nomogram by painting a grid onto a piece of wood. Hammer small nails to the integer coordinates for $Y=X^2$. Use yarn to connect each coordinate on the right of the Y-axis to each coordinate on the left of the Y-axis to make your own multiplication chart (and string art!).

Chapter 5 (answers on page 274)

Topic 5.1: Camelot

Camelot is a castle found in the British folklore of King Arthur and the Knights of the Round Table. Arthurian Legends are stories that were developed over hundreds of years by different authors. They all center around the character of King Arthur and share the common themes of adventure, love, and chivalry.

In "*Math and Magic in Camelot*", you'll find some references to traditional Arthurian legends as well as the Celtic mythology from which they originated. Since this book takes place AA ("After Arthur"), however, the ladies of the court will have a chance to shape their own narrative of honor and bravery.

To read more of the traditional tales of King Arthur and the Knights of the Round Table, look for this book at your local library:

> *King Arthur and the Legends of Camelot*
> Molly Perham
> 1993

Topic 5.2: Three is a Magic Number

To the Lily Maidens, the three petals of a lily represent the "three-petaled way" – the code of honor by which the Ladies live (although its details are not revealed in this chapter).   Finding spiritual meaning in the number three is not a new concept. The Latin phrase "omne trinum perfectum" means that everything that comes in threes is perfect. What is so perfect about the number three?

Use an internet search engine to find a video (or lyrics) for the song *"Three is a Magic Number"* written by Bob Dorough for the *"Schoolhouse Rock"* series.

How many examples of the "magic" of the number three can you find? Make a list! Here are a couple of ideas to get you started:

- The trefoil knot with its three intersections is the first non-trivial knot. (You already know this!)

- Three is the first odd prime number.

- Many fairy tales have three parts or three main characters – *"The Three Little Pigs"*, *"The Three Billy Goats Gruff"*, *"Goldilocks and the Three Bears"*.

- Three points define a plane; A three-legged stool will not tip; Construction with triangles is the most stable.

- The sum of all the digits of a multiple of three will also be divisible by three. Example: 534 is divisible by three, 5+3+4 = 12, which is also divisible by 3, and 1+2=3.

What additional examples can you find?

Topic 5.3: Parts of a Flower

Upon their arrival in Camelot, Lulu and Elizabeth were immediately presented with a botany question: *How many petals has the lily?* Fortunately for the girls, Elizabeth was already familiar with the correct names for the parts of a flower and answered the question correctly (a lily has three petals and three sepals).

Purchase a lily (or another large flower like a daffodil or tulip) from your local florist and see if you can identify the parts of the flower labeled in the diagram below. To see the ovules, you'll have to gently pull the flower apart or cut it open with a plastic knife (adult supervision required).

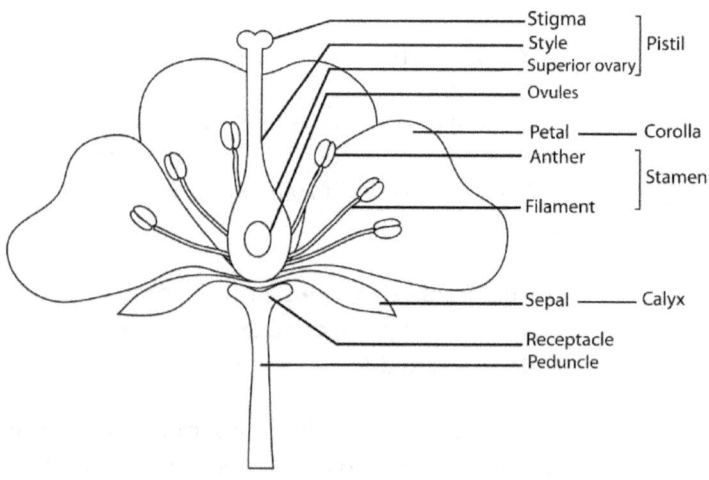

Here are the main flower parts to know:

- Petals give a flower its shape and color. They attract pollinators such as bees and butterflies to the flower.

- Sepals enclose and protect the flower bud as it develops. They remain attached to the base of the flower after it opens. In most flowers, the sepals are green, but in many lilies, they are the same color and size as the petals.

- The stamen is the male part of the flower. A thin stalk called the filament supports the roundish pollen-producing anther. Bees and other insects will carry the pollen from flower to flower.

- The pistil is the female part of the flower. The sticky stigma catches pollen brought by insects from other flowers. The pollen will travel down a tube called a style to the ovary, where it pollinates the little round ovules to form seeds.

Which parts of your lily come in multiples of three?

Topic 5.4: The Rose Family

Part A

Read the poem *"The Rose Family"* by Robert Frost (published in his 1928 collection *"West-Running Brook"*). Perhaps Frost was as confused as Lulu by the botanical classification of apples, pears, and plums as part of the rose family (all having five sepals).

Modern-day biologists use genetic tests to identify how different species of plants and animals are related to each other for accurate classification. Similarly, in object-oriented software architecture, computer scientists identify relationships between various types of objects in their software programs.

Fill in each sentence to indicate either a compositional relationship ("has a") or a taxonomy ("is a"):

Fill in with:
A: "has a"
B: "is a"

1. A flower _____ petal.
2. A rose _____ flower.
3. A pear _____ rose.
4. A pistil _____ stigma.
5. A stamen _____ filament.
6. A plum _____ rose.

Can you find a way to draw these relationships?

Topic 5.5: Base Two

To activate the traveling bed with the flowered quilt, Lulu had to enter the number of sepals for the destination using base-two.

The numeral system we use most frequently is the Hindu-Arabic system, which has ten numerals – from zero to nine. This system is also called base-ten.

On the headboard of the bed with the flowered quilt, there are buttons which have only two states – flower open or flower closed (on or off). In base-two systems (also called binary systems), you must write numbers using only two digits – zero and one.

<u>Part A:</u>

If the headboard only contained one button, Lulu would only be able to enter two values – zero (flower closed) and one (flower open). With two buttons, she could enter four values:

Button 2	Button 1	Binary (Base-2)	Value (Base-10)
●	●	00	0
●	✿	01	1
✿	●	10	2
✿	✿	11	3

In base-two, 10 is not "ten", but a one and then a zero, which equals a value of 2 in the Hindu-Arabic numeral system (base 10) that most people are familiar with. Here's a classic math joke to baffle your friends: *"There are only 10 types of people in the world: those who understand binary, and those who don't."*

If the headboard has three buttons, how many values can be represented? Copy the chart below onto your own piece of paper and complete the last two rows:

Button 3	Button 2	Button 1	Binary Base-2	Value Base-10
●	●	●	000	0
●	●	 ✿	001	1
●	 ✿	●	010	2
●	 ✿	 ✿	011	3
 ✿	●	●	100	4
 ✿	●	 ✿	101	5

Did you find the pattern for binary counting? What is the highest number you could represent using four on/off buttons? What about five buttons?

<u>Part B:</u>

How did Lulu convert the number eighty-five to binary (base two)? First, she wrote the following numbers in her notebook:

64, 32, 16, 8, 4, 2, 1.

As you discovered in part 1, one button can represent the integers 0 or 1, the second button adds two more values (for a total of 4 options), the third adds four new values (for a total of 8 options), the fourth adds eight new values (for a total of 16 options), and so on. If Lulu had continued the pattern, she could have written 128 to the left of the number 64.

To convert 85 from base-ten to binary, Lulu put "1" underneath the 64 since 64 goes into 85 one time. Next, she wanted to see what was left: 85-64=21. Since 32 is larger than 21, she placed a zero below it. Continuing to the right, Lulu placed a 1 below 16 because 16 goes into 21.

64	32	16	8	4	2	1
1	0	1				

What is left? 21-16=5, which is zero eights and one four, with one remaining:

64	32	16	8	4	2	1
1	0	1	0	1	0	1

To check your answer, add 64+16+4+1=85.

Try converting these numbers into binary (base two):

A. Eli is eight years old. What is his age in base-2?

B. "Thirty days hath September." Write 30 in base-2.

C. There are 52 weeks in a year. Write 52 in base-2.

Chapter 6 (answers on page 276)

Topic 6.1: William Butler Yeats

William Butler Yeats (1865-1939) was an Irish author who retold many of his country's myths and legends in the form of poetry. In this chapter, Yeats's poem *"The Song of Wandering Aengus"* is presented not as the tale of Angus, the Celtic god of love, but as the narrative of Yeats's encounter with the enchanting Nimue. William Butler Yeats was known as a man who believed in fairies, ghosts, and other supernatural phenomena, so picturing him in this fictional role does not take a far stretch of the imagination.

To read the traditional Irish tale on which *"The Song of Wandering Aengus"* was based, look for this beautifully illustrated book at your local library:

> *The Dream of Aengus*
> By Joanne Findon and Ted Nasmith, 2015

You can read more of the poems written by William Butler Yeats in this collection:

> *William Butler Yeats*
> *(Poetry for Young People Series)*
> By W. B. Yeats, Jonathan Allison, and Glenn Harrington, 2002.

Topic 6.2: How Crystals and Gemstones Form

Part A

Only in folklore can a fruit transform into a diamond. Natural diamonds are carbon crystals that form under very high temperature and pressure in the Earth's mantle over billions of years.

The carbon atoms in a diamond are arranged into a cubic structure called a diamond lattice. On the other hand, graphite (the material you find in pencil "lead") is made of the same element as a diamond (carbon), but its atoms are arranged in layers (planes).

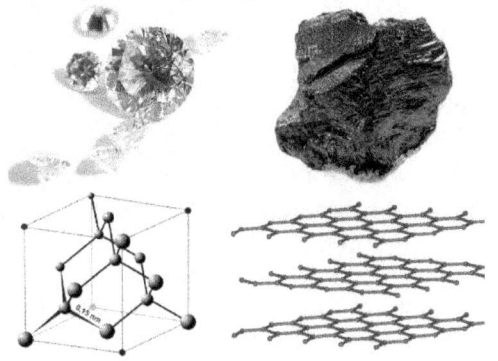

60

Diamonds are very hard and strong. They vary in color from clear (colorless) to various shades of yellow, brown, or other colors. Examine the "lead" of your pencil and list some of the properties of graphite. How do the properties of graphite differ from the properties of a diamond?

<u>Part B</u>

You can grow sucrose (sugar) crystals at home. You'll need 3 cups of water in a microwave-safe container, one cup of sugar, a clean glass jar, a pencil, and a piece of yarn. Directions:

1. Make sure that the pencil can sit across the jar opening without falling in. Tie the yarn to the pencil and cut the end of the yarn so that it dangles near the bottom of the jar.

2. Ask an adult to boil the water in the microwave. Microwave times will vary so the water must be checked frequently.

3. Carefully add one teaspoon of sugar to the water and stir until it is dissolved. Repeat the process, adding one teaspoon of sugar at a time and stirring. Stop when you see that the sugar is no longer dissolving.

4. Ask an adult to pour the hot sugar solution into your glass jar.

5. Place the pencil across the opening of your jar so that the yarn dangles from its center into the solution of dissolved sugar.

6. Leave your jar in a location where it can sit undisturbed. You may place a piece of paper towel over the top of the jar so dust doesn't fall inside.

7. Check on your jar every day to watch the sugar crystals form. After a few days, pull out the yarn and examine the sugar crystals with a magnifying

glass. What shape are the crystals? Draw them. If you used a clean glass jar and clean yarn and did not put anything other than water and sugar into your container, you can eat your "rock candy"!

Topic 6.3: Missing Gem Puzzle

The Fairy Queen claimed that she had created two new jewels which Nimue had lost or stolen. See if you can find the error in her logic:

1. Nimue brought 20 gems to the Fairy Queen for safe keeping.

2. At the end of the first week, Nimue retrieved 6 gems, leaving 14 (20-6=14).

3. At the end of the second week, Nimue retrieved 6 additional gems, leaving 8 (14-6=8).

4. At the end of the third week, Nimue retrieved the remaining 8 gems. None were left.

5. The Fairy Queen claimed that she had created two extra gems – after all, there were 14 gems left after the first week and 8 gems left after the second, for a total of 22. Nimue only had 20 gems in her possession. Where did the other 2 gems go?

Chapter 7 (answers on page 277)

Topic 7.1: Pigeonhole Principle

    The mathematical concept called the "pigeonhole principle" is very intuitive: If you have more containers than objects, then at least two objects will have to share a container.

    For example, if you have ten pigeons and nine pigeonholes, then at least one hole must contain more than one pigeon.

    In this chapter, Elizabeth and Lynette need two pieces of fruit of the same color from the fruit-dispensing machine. There are three different colors of fruit, so imagine three "pigeon holes", each painted with a different color in which the fruit gets sorted. For a matching pair, you want to guarantee that two pieces of fruit share the same color ("pigeon hole"). Using the pigeon-hole principle, Elizabeth reasoned that with four pieces of fruit, you would be guaranteed that at least two of them will be the same color.

Use the pigeon-hole principle to solve these additional problems:

A. Imagine a drawer full of socks – half of them green and the other half blue. If you reach into the drawer at night (in the dark), how many socks should you pull out to guarantee a matching pair?

B. In the example above, how many socks should you pull out to guarantee two matching pairs?

C. What is the minimum number of people you should invite to your party to be sure that at least two of them will share a birthday?

D. Can each person at your party shake hands with a unique number of people? For example, the first person doesn't shake hands with anyone; the second person shakes hands with one person, the third shakes hands with two different people, and so on. Is this possible?

Topic 7.2: Emily Dickinson

American poet Emily Dickinson (1830-1886) is Elizabeth's favorite. Elizabeth especially enjoys the nature themes found throughout Dickinson's poetry.

Emily Dickinson's interest in botany began at an early age as she took care of the gardens at her family's homestead. Over her lifetime, Dickinson collected and identified over 400 plant specimens which she pressed into an herbarium.

In her later years, Emily Dickinson lived a reclusive life, never leaving her home. She corresponded with friends only through letters but was said to enjoy the company of children.

This charming picture book tells a (fictionalized) story of a child meeting Emily Dickinson:

Emily
By Michael Bedard
Illustrated by Barbara Cooney
2008

Emily Dickinson wrote over 1,800 poems, nearly all of which were published only after her death. For more of Emily Dickinson's poetry, look for this illustrated collection:

Poetry for Young People: Emily Dickinson
By Emily Dickinson
Illustrated by Chi Chung
2014

Topic 7.3: The Ballad of the White Horse

In this chapter, Elizabeth learns that King Arthur was inspired by Chesterton's epic poem *"The Battle of the White Horse"* when he ordered his own White Horse Hill. Chesterton's poem tells the story of Saxon King Alfred the Great as he fought the invading Danes. Ironically, King Arthur, although a mythical character, is also regarded as a historical figure who led the Britons against the Saxon invasion. Would Arthur have resented the poem had he known its subject, or would he sympathize with Alfred's struggle to protect his homeland?

Look for this illustrated edition of Chesterton's poem:

The Ballad of the White Horse
By G.K. Chesterton
Illustrated by Ben Hatke
2011

For a fun nature craft, collect white or light-colored stones and arrange them in a grassy area to form your own "white horse" (check with an adult for an acceptable location). Don't forget to scour your horse frequently!

Topic 7.4: Celtic Mythology – Finn McCool

Fionn mac Cumhaill ("Finn McCool") is a great hero of Celtic mythology. For more of Fionn's story, along with other classic Irish folklore, look for this book in your local library:

> *Traditional Irish Fairy Tales*
> By James Stephens
> Illustrated byArthur Rackham
> 1920

Topic 7.5: Gregor Mendel and the Genetics of Peas

Elizabeth speculated that the color of the rowan fruit is a trait that passes on to the next generation following the laws of Mendelian inherence. Gregor Mendel (1822 - 1884), a monk and scientist, is known as the father of modern genetics. His theories on inheritance were developed through experimentation with pea traits. Look for this book at your local library for an excellent introduction to the life of Gregor Mendel and his research with peas:

Gregor Mendel: The Friar Who Grew Peas
By Cheryl Bardoe
2015

As Elizabeth explained to Lynette, as the first act of his "magic trick", Mendel bred plants that had different expressions of a trait – for example, round peas and wrinkled peas. When a plant with round peas was bred with a plant with wrinkled peas, all their offspring had round peas:

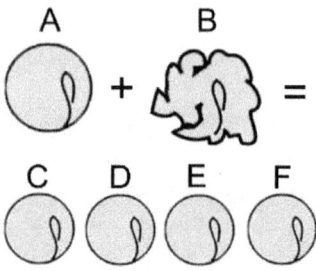

In what Elizabeth described as the "Second Act" of the "magic trick", Mendel bred the plants in this new generation with each other. While most of the resulting offspring had round peas like their parents, a few of the offspring yielded wrinkled peas (like their 'grandparents'). Mendel did not just record that "some" of the offspring were wrinkled. Instead, he looked at the results of his experiment like a mathematician and determined what percentage of the plants (one-fourth) were wrinkled rather than round. Mendel's statistical analysis helped uncover the biology of genetic inheritance.

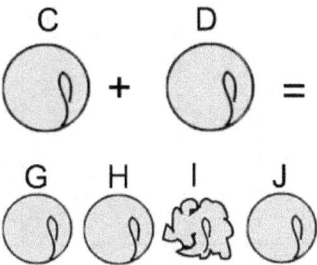

C   D

G   H   I   J

Do you remember the parts of a flower from the Chapter 5 activities? The pollen from a plant contains genes - instructions for different traits. Likewise, the egg in a plant's ovule also contains genes. When fertilization takes place, a seed will grow with genes (instructions) from both parent plants.

"Round peas" is a <u>dominant trait,</u> which means that a plant only requires one gene for "round" to have round peas (even if the other gene is "wrinkled"). "Wrinkled peas" is a <u>recessive trait,</u> so both genes must be "wrinkled" for the plant to have wrinkled peas. We use upper-case letters to represent dominant traits and lower-case letters for recessive traits.

R = round allele
r = wrinkled allele

RR = round, Rr = round, rr = wrinkled

A Punnett square is a diagram that shows all the different possible trait results. If a pure round pea plant is bred with a wrinkled pea plant (like in the first part of Mendel's experiment), this is the result:

		Pure Wrinkled (rr)	
		r	r
Pure Round (RR)	**R**	Rr	Rr
	R	Rr	Rr

As you see in the diagram, all the offspring carry a gene for round peas and a gene for wrinkled peas, but they only display the dominant trait (round).

Question A:

If you breed two of these new plants, what are the possible results? Copy the Punnett square below on a separate paper and complete it. Draw either a round pea or a wrinkled pea next to each resulting plant (remember RR = round, Rr = round, rr = wrinkled). What percent of the child plants are expected to yield wrinkled peas?

		Heterozygous Round (Rr)	
		R	**r**
Heterozygous Round (Rr)	**R**		
	r		

Question B:

Mendel studied seven different traits of the pea plant: seed shape, seed color, flower color, pod shape, pod color, flower position, and stem height. If each trait has two possible values (shape is either round or wrinkled, color is either yellow or green, stem height is either tall or short, etc...), how many different combinations of these characteristics exist?

Hint: Start with one trait and keep adding traits to find the total number of combinations. Look for a pattern.

Question C:
The trait for yellow peas is dominant. A plant of type YY or Yy will produce yellow peas, and only a plant of type yy will have green peas. We already know that round (R) is dominant over wrinkled (r). Let's look at these two traits together.

If you breed a pure yellow, round pea (YYRR) plant with a green, wrinkled pea plant (yyrr), what will be the color and shape of the resulting peas? Complete the Punnett square:

		Yellow & Round (YYRR)			
		YR	**YR**	**YR**	**YR**
Green and Wrinkled (yyrr)	**yr**				
	yr				
	yr				
	yr				

Question D:
If you breed two of the offspring from the second generation in the last question, what traits will the third generation exhibit? Complete the Punnett square:

		YyRr			
		YR	**Yr**	**yR**	**yr**
YyRr	**YR**				
	Yr				
	yR				
	yr				

Out of the sixteen cells in your table, count how many were:

Yellow and Round: ____ / 16
Yellow and Wrinkled: ____ / 16
Green and Round: ____ / 16
Green and Wrinkled: ____ / 16

Chapter 8

Topic 8.1: Dicots, Monocots, and Victoria

Part A

In this chapter, Lulu discovered that Victoria is a water lily. Water lilies are members of a group called Nymphaeaceae and are not related to the "true lilies" of the Lilium family.

The Victoria water lily (named after Queen Victoria) grows in the Amazon River. Its huge circular leaves, which float on top of the water, can grow almost nine feet across. A mature leaf can support as much as 100 lbs. of weight, so it is plausible that Lulu could stand on Victoria without sinking.

Victoria's stalk can stretch 20 feet down through the water. What do you think Emily Dickinson meant when she asked if Morning has *feet like water lilies*?

Victoria's large flowers are approximately 15 inches across. On the first night, a Victoria flower is white. It is pollinated by the scarab beetle who is attracted to the Victoria's sweet fragrance. The beetle is trapped inside the flower overnight, where the pollen that is stuck to its body from visits with other Victoria water lilies is rubbed onto the stigma.

On the second night, Vitoria will change from female to male, producing pollen for the beetle to carry to other water lilies after it is released. The Victoria flower will turn pink. When the second night is over, the job of fertilization has been completed, and the flower will sink beneath the water.

The Victoria water lily is depicted on this Hungarian postage stamp:

62

Part B

Lady Olwyn suggested that the ladies do not completely trust Victoria because she is a dicot. Monocots like the true lily have only a single initial leaf or cotyledon. Grasses also fall into this category. Most

---

plant species are dicots, which means that they have two cotyledons.

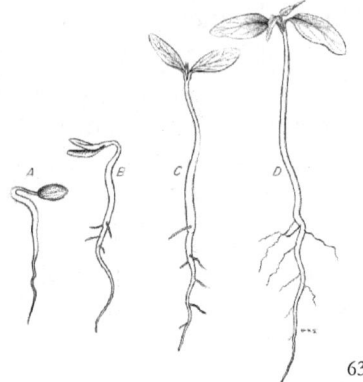

Explore the plant structure of monocots and dicots:

1. Line a disposable plastic cup (or another container) with sheets of paper towel.

2. Place a couple of dry beans inside and add enough water to get the paper towel soaked.

3. Leave the cup undisturbed and check on your beans daily. Add water as needed, so the paper towel remains moist.

4. After a couple of days, you should see your beans germinate. Roots will soon appear, followed by the embryonic leaves. How many of these first leaves do you see – one or two? Is a bean a monocot or dicot?

---

63 By W.H.L. @ USDA-NRCS PLANTS Database [Public domain], via Wikimedia Commons

5. Compare the structure of your bean leaves to the structure of grass leaves (or lily leaves). Draw a diagram of each. How are they different?

6. Dig up grass roots and compare their structure to those of your bean plant roots.

Topic 8.2: Arthurian Legend: The Loathly Lady

"*The Weddynge of Syr Gawen and Dame Ragnell*" is an English poem written by an unknown author in the 15th century. It retells the popular story of a 'loathly lady'– an ugly woman who is transformed when a man looks past her physical features.

The poem is written in Middle English (spoken between the 12th and 16th centuries). Here is the beginning of "*The Weddynge of Syr Gawen and Dame Ragnell*".

> *Lythe and lystenyth the lif of a lord riche!*
> *The while that he lyvid was none hym liche,*
> *Nether in bowre ne in halle.*
> *In the tyme of Arthoure thys adventure betyd;*
> *And of the greatt adventure, that he hymself dyd,*
> *That Kyng curteys and royall.*

See if you can decipher the words you don't recognize by sounding them out, looking at their context, and comparing them to modern-day English spellings. Here is a translation of the passage above to modern English:

> *Pay attention and listen to a tale of the life of a wealthy lord! While he was alive, there was no one like him anywhere in the land. It was in the time of King Arthur*

that this adventure unfolded, for it is a story of the courteous and royal King himself.

Look for this illustrated retelling of the Arthurian legend (in modern English) at your library:

<u>*Sir Gawain and the Loathly Lady*</u>
By Selina Hastings
Illustrated by Juan Wijngaard
1988

Topic 8.3: Buoyancy – Floating Egg Experiment

The Fairy Queen revealed to Lulu that Dozmary Pool, also known as the "Lake of Tears", is getting saltier. The magic gems which normally lie on the bottom of the lake under the protection of the water lily's feet have started to float.

Why do some objects sink and others float? Buoyancy – an object's tendency to sink or float – is determined by the downward force of the object (its weight) compared to the upward buoyant force (the amount of water that it displaces).

When salt dissolves in freshwater, the water molecules and salt molecules will pack tightly together, so the water doesn't increase in volume, but will increase in weight. Since saltwater is heavier than freshwater of the same volume, an object will displace a smaller volume of saltwater than freshwater, and it will be more buoyant. If the water gets salty enough, an object that previously sank to the bottom will begin to float.

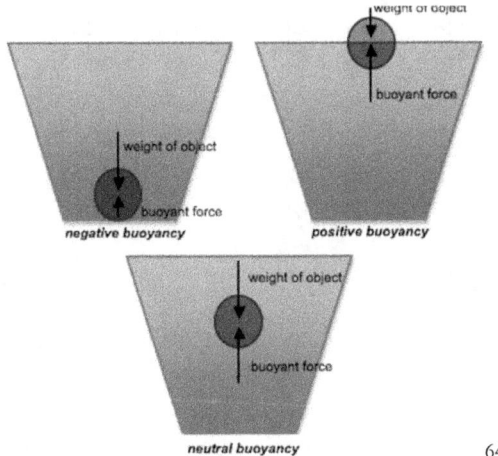

weight of object

buoyant force

negative buoyancy

weight of object

buoyant force

positive buoyancy

weight of object

buoyant force

neutral buoyancy

64

Can you make an egg float in salt water? Try the following:

1. Gather your materials: A clear cup or bowl which can hold at least three cups of water, a tablespoon (measuring spoon), a different spoon for stirring, table salt, a raw egg.

2. Fill your cup or bowl with 1 cup of water. Add 4 tablespoons (1/4 of a cup) of salt to the water and begin stirring. Stir until all the salt is completely dissolved. This may take as long as ten minutes of continuous stirring.

3. Place the raw egg in your salt solution. The egg should float! This is a state of positive buoyancy.

4. Add ¼ cup of fresh water to your solution and stir gently. Keep diluting your salt-water by adding fresh water ¼ cup at a time and observing

64 By Khursheed afroz, licensed under CC BY-SA 4.0

the egg. How much fresh water did you add before the egg reached neutral buoyancy? How much fresh water before reaching negative buoyancy?

Chapter 9 (answers on page 278)

Topic 9.1: Arthurian Legend: Sir Yvain and the Lion

The Pool of Joy in this story is inspired by an Arthurian legend of a magical spring that bubbles as if boiling, yet is *cold as marble*. When water from the spring is poured onto a stone, a massive storm is immediately produced. After hearing an account of the magical spring, Yvain is determined to see it for himself. Thus, his adventure begins.

"Yvain, the Knight of the Lion", was written by French poet Chrétien de Troyes in the 12th century and tells the tale of Yvain and the magical spring. The story is retold in this illustrated book:

> *The Knight with the Lion: The Story of Yvain*
> By John Howe
> 1996

A modern retelling of the Arthurian tale in graphic-novel form is also available:

> *Yvain: The Knight of the Lion*
> By M. T. Anderson
> Illustrated by Andrea Offermann
> 2017

Topic 9.2: Fill and empty container puzzles

Lulu has two basins – a larger one that holds 5 units and a smaller one that holds 3 units. How can she collect exactly 4 units of water by filling and emptying the two basins?

(Try to solve the problem yourself before reading on.)

Lulu already completed the first couple of steps. She filled the 5-unit basin with water and poured it to fill the 3-unit basin. There were 2 units left in the larger basin (5-3=2). Next, she poured the water from the smaller basin back into the pool.

If she hadn't been interrupted by the Fairy Queen, Lulu would have poured the 2 units of water from the larger basin into the smaller one. Now the smaller basin only has room for 1 unit of water. Lulu would have to refill the 5-unit basin and use it to completely fill the smaller one. Since 5-1=4, the larger basin would have the exact volume of water needed to 'dilute the sorrow' from the Lake of Tears.

There are a whole group of math puzzles where you are asked to fill and empty containers to achieve a specified volume of water (or other substance). Here is another one for you to try:

A recipe calls for 5 cups of flour. How can you pour the exact amount of flour into your mixing bowl using only a 4-cup measuring cup and a 7-cup measuring cup? (Note: you're free to pour flour back into the bag of flour after filling the measuring cups.)

Topic 9.3: Buffon's Needle – Finding π (pi)

If you drop a pin onto a striped bedspread, what is the chance that it will land in-between two stripes? In the 18th century, mathematician Georges-Louis Leclerc, Comte de Buffon posed a similar question (but involving a needle dropped on a floor made of parallel wood strips). Lulu recalled the story and thought of the number π (called 'pi').

Reciting the first one-hundred digits of pi helps Lulu relax, which explains how she knew in what order to place the numbered marbles in the headboard of the bed. Pi is a special number that describes the ratio of a circle's circumference (the distance around it) to its diameter (the distance across it, passing through the center).

How is this related to Buffon's question? Try this experiment:

1. Instead of using pins or needles, find a toothpick (or a match with the head cut off, a drinking straw, a pencil, or a stick).

2. Using a ruler, draw parallel lines on a piece of paper so that the distance between the lines is exactly one toothpick-length (or the length of whichever object you've chosen).

3. Place the paper on a table and drop the toothpick onto the paper from a height of approximately 6 inches (adjust the drop height if your toothpick doesn't land on the paper. On a separate paper, record whether your toothpick landed:

A. Not touching any line.
B. Touching or crossing a line.

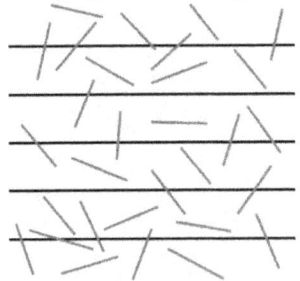

4. Repeat Step 3 one hundred times. Use tally marks to record the results.

5. Out of the 100 trials, how many times did the toothpick touch or cross a line (B)?

6. Using a calculator, compute:

$200 \div B$

You should get an approximation of $\pi = 3.14159\ldots$ (and *that's* magic!)

Topic 9.4: Morse Code

Lulu guessed that the pattern of dots and dashes around the smaller basin was Morse code. Morse code, named after the inventor of the telegraph, Samuel F. B. Morse, is an international message system. Letters (and numbers) can be transmitted using dots and dashes (binary!).

To decode a letter using the Morse code tree below, begin at "start" and follow the direction of dots and dashes. For example, to decode — ●●●, put your finger on "start", then move to the right (dash) once and then to the left (dot) three times until your finger lands on the letter "B".

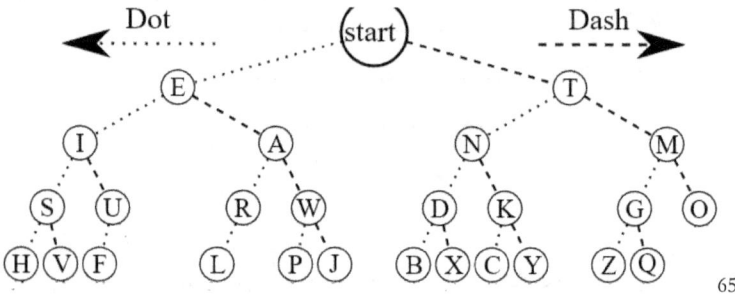

Use the Morse code tree to decypher the message from the smaller basin:

B					
— ●●●	●	● — —	● —	● — ●	●

Make a necklace of your name in Morse code. Use round beads for dots and elongated beads for dashes. You could also form beads out of clay or make beads by cutting plastic drinking straws.

Topic 9.5: Brachistochrone curve and cycloids

Look at the picture below. If two marbles are dropped from point A at the same time – one on the straight track and the other on the curved path – which one will arrive at point B first?

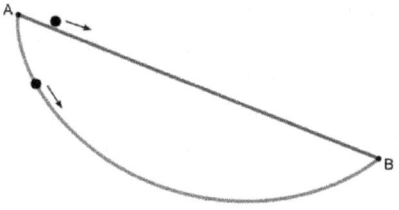

Lulu was confident that since the straight line from A to B is the shortest path, a marble (or gem) would travel along the straight track in the shortest amount of time. She was wrong! The curved track (called a brachistochrone curve) is the fastest path from A to B.

You can test out the brachistochrone curve for yourself. Use flexible toy racecar tracks (such as those from Hot Wheels™ sets) or make your own tracks out of foam pool noodles (cut them in half lengthwise).

The curve you need is called a cycloid. It is formed by a point on a wheel rim as it travels in a straight line. A cycloid is a special form of a "trochoid curve". If you've ever played with a Spirograph toy, you are already familiar with trochoid art!

Try drawing a cycloid by taping a marker to the inside rim of a jar lid so that the marker is perpendicular to the lid. Tape a piece of paper to the wall and roll the jar lid in a straight line across the floor as the marker draws on the paper (don't draw on your wall!). You should get a curve that looks like this:

CYCLOID

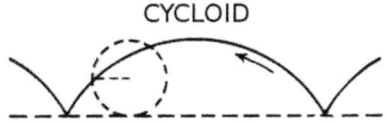

To create the template for the brachistochrone curve, you will need a much larger circular object (like a large circular laundry basket, hula hoop, or huge bowl). Attach the marker to this larger circle and set poster board against the wall. Roll the circular object to draw your cycloid.

Cut out the cycloid from cusp to cusp (the point where the curve is done descending and begins to rise again) and tilt it at any angle you desire (as long as one end is lower than the other). Position one flexible track along this cycloid template. You may need to tape the track in position or use blocks beneath the track to keep it in place. Position the other track so it descends in a straight line. Place the two tracks side by side, so they look like this:

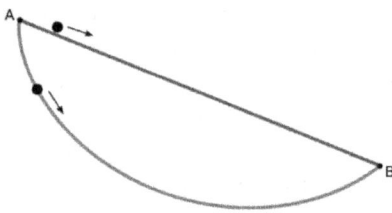

Drop two marbles at the same time (one in each track) to see which path results in the shortest travel-time.

Chapter 10 (answers on page 279)

Topic 10.1: Chinese Remainder Problem

Elizabeth and Lynette overheard some men discussing the number of troops that were already gathered at Black Horse Vale. Here is the information they gathered:

1. Each unit has 100 men.

2. When the units lined up with 3 units in each row, 2 units were left over.

3. When the units lined up with 7 units in each row, 3 units were left over.

4. When the units lined up with 10 units in each row, none were left over.

Using this information, how many men were in the valley? (Keep reading for the answer)

You can use a hundreds chart to narrow down the options. Print your own hundreds chart (or make one out of graphing paper) and follow along.

Since there were no units left when they lined up by 10s, you know that the number of units must be a multiple of 10. Color all the numbers in column J:

A	B	C	D	E	F	G	H	I	J
1	2	3	4	5	6	7	8	9	10
11	12	13	14	15	16	17	18	19	20
21	22	23	24	25	26	27	28	29	30
31	32	33	34	35	36	37	38	39	40
41	42	43	44	45	46	77	48	49	50
51	52	53	54	55	56	57	58	59	60
61	62	63	74	75	66	67	68	69	70
71	72	73	74	75	76	77	78	79	80
81	82	83	84	85	86	87	88	89	90
91	92	93	94	95	96	97	98	99	100

When they lined up 3 units to a row, there were 2 units left over. Find each multiple of 3 and add 2 (for the two remaining units). 0+2=2, 3+2=5, 6+2=8, etc… You can skip count by 3s to fill in the chart:

A	B	C	D	E	F	G	H	I	J
1	2	3	4	5	6	7	8	9	10
11	12	13	14	15	16	17	18	19	20
21	22	23	24	25	26	27	28	29	30
31	32	33	34	35	36	37	38	39	40
41	42	43	44	45	46	77	48	49	50
51	52	53	54	55	56	57	58	59	60
61	62	63	74	75	66	67	68	69	70
71	72	73	74	75	76	77	78	79	80
81	82	83	84	85	86	87	88	89	90
91	92	93	94	95	96	97	98	99	100

The numbers 20, 50, and 80 are colored in both charts (so they meet the conditions of being both multiples of 10 and having a remainder of 2 when divided by 3).

When the troops were divided into rows of 7, there were 3 troops left over. Use the same process to color the numbers that are 3 more than each multiple of 7:

A	B	C	D	E	F	G	H	I	J
1	2	3	4	5	6	7	8	9	10
11	12	13	14	15	16	17	18	19	20
21	22	23	24	25	26	27	28	29	30
31	32	33	34	35	36	37	38	39	40
41	42	43	44	45	46	77	48	49	50
51	52	53	54	55	56	57	58	59	60
61	62	63	74	75	66	67	68	69	70
71	72	73	74	75	76	77	78	79	80
81	82	83	84	85	86	87	88	89	90
91	92	93	94	95	96	97	98	99	100

The only integer that fits all <u>three</u> conditions is 80, so this is the smallest number of troops at Black Horse Vale. There are 100 men to a troop, so there must at least $80 \times 100 = 8,000$ men waiting to attack Camelot.

This type of puzzle is called a Chinese Remainder Problem as it first appeared in *"Sunzi Suanjing"*, a 3rd-century Chinese text on mathematics. Here is another one for you to try:

A woman arrived at the market with a basket full of eggs. She placed the basket down and surveyed the busy market. Just then, a horse appeared from around the corner and stomped right on the basket of eggs. What a mess! The horse's owner apologized profusely and offered to pay the woman for her loss.

"I don't know exactly how many eggs were in the basket," admitted the woman. "But I did move them into a couple of different baskets before I found the perfect one," she recalled. The woman looked down with a sigh at her favorite pink basket that was now covered in yolks and shells. "When I took them out 3 at a time, 2 eggs were left in the basket. Next, I took them out 5 at a time, and there were 3 left. Finally, I took the eggs out 7 at a time. As you see, I have large hands which are rather useful for egg-handling," the woman said and proudly displayed her thick sunbaked hands.

"Seven at a time. I see," the man responded, feigning interest.

"Yes, that was how I transferred the eggs to the final basket. And there were 2 left."

The woman studied the man's face. He nodded his head, but his eyes were distant, thinking. After several minutes of silence, the man pulled some coins out of his satchel and paid the woman for her eggs.

Based on the information she provided, what was the minimum number of eggs the woman could have had in her basket?

Topic 10.2: Aspen Grove as a Single Organism

What is the largest living organism on Earth? If you think the title belongs to the blue whale, you're wrong. Although the blue whale (who can weigh up to 190 tons) is the largest animal on our planet, the Aspen colony is the largest organism.

Above the surface, an Aspen (Populus) Grove may appear as a large group of separate trees. But below the surface, a massive root system reveals that all the trees in the grove are connected. They are genetically identical - all part of the same organism: a clonal colony. You can see why the Lily Maidens in the story have chosen to send messages by Populus Post.

Pando, the "Trembling Giant" is a Quaking Aspen grove that is considered the largest living organism on our planet. Scientists estimate Pando's weight at 6,600 tons. Pando also holds the title for the oldest living organism at approximately 80,000 years old.

67

The petiole is the name for the part of a tree which attaches a leaf to the stem. In a quaking aspen (populus tremuloides), these petioles are flat, which causes the leaves to shake in even the slightest breeze. Now you know why the aspen quakes!

67 By Carl Axel Magnus Lindman [Public domain], via Wikimedia Commons

Topic 10.3: Arthurian Legend – Merlin

Merlin the wizard plays a primary role in many Arthurian legends as King Arthur's trusted advisor. Many authors have added details to Merlin's character over the years in their adaptations of the popular myths. This book uses Sir Thomas Malory's treatment of Merlin as the devil's son. Nimue, one of the Ladies of the Lake, is also a part of Mallory's "*Le Morte d'Arthur*" (first published in 1485). Unlike many of the female characters in Arthurian literature, Nimue is not merely a weak damsel or temptress, but a strong figure in her own right.

For more stories of King Arthur and Merlin, look for these titles:

Merlin and the Dragons
By Jane Yolen
Illustrated by Li Ming
2009

Merlin and the Making of the King
Retold by Margaret Hodges
Illustrated by Trina S. Hyman
2004

Chapter 11 (answers on page 279)

Topic 11.1: Toothpick Puzzles

Move two pillars and not one more,
To release the Wizard and open the door.

Were you able to successfully solve the problem along with Elizabeth? Her solution looked something like this:

1.　　　**2.**　　　**3.**

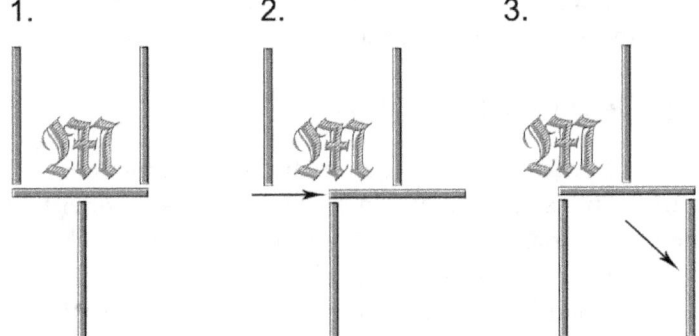

These types of brainteasers are called "toothpick puzzles" (sometimes "matchstick puzzles") because they involve placing toothpicks into a particular formation and then moving the specified number of toothpicks to make a new configuration.

Give this toothpick riddle a try:

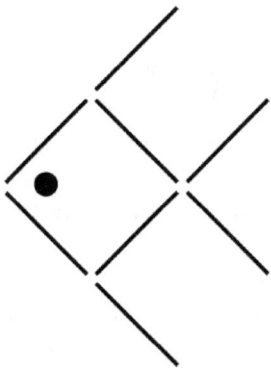

Arrange your toothpicks to copy the fish formation above. *Could it be Nimue as the "Fish of Desire"?*

Move three toothpicks (and the eye) to make the fish swim in the opposite direction.

Topic 11.2: Friendly Numbers

Lynette told Elizabeth that the traveling beds come in pairs and the one in Merlin's cave is number 84 - the friendly pair with bed #270 (the one upon which Lulu and Elizabeth traveled to Camelot).

Friendly pairs are found in number theory -the study of integers, or whole numbers (not fractions). Two (or more) numbers are "friendly" if they have the same abundancy. Here's how abundancy is calculated:

1. Find all the divisors of your number.

2. Add up all the divisors you found.

3. Divide the sum of the divisors by the number itself to get the abundancy.

For example, take the number 6. It can be divided by 1, 2, 3, and 6. Since $1+2+3+6=12$, its abundancy is $12÷6=2$. Mathematicians have a name for this – six is called a "perfect" number.

The number 28 can be divided by 1, 2, 4, 7, 14, and 28. The sum of those divisors is 56. With an abundancy of 2, it is also a "perfect" number and a friendly pair with 6.

Friendly numbers don't just come in pairs. The numbers 84 and 270, for example, have the same abundancy as 1488, 1638, and 24384. Do you suppose there are additional gigantic traveling beds with those numbers?

Numbers with no friends are called "solitary". Mathematicians think that 10 is a solitary number, but they are still trying to prove it.

You probably won't find many practical uses for friendly numbers in your everyday life. Mathematicians simply love to play with numbers and discover new patterns and relationships.

Give it a try: Use twelve blocks or square crackers to find all the divisors of the number 12. If you can arrange the twelve items into a rectangle with N rows and no items left over, then N is a divisor of 12. Add up the divisors and then divide the sum by 12 to get the number's abundancy.

Topic 11.3: Mythological Monsters

Merlin's "device" appears to use the blood of a creature to bring it back to life. Who are the three mythological giants that have been conjured?

Grendel

Grendel is a man-eating monster from the Old English epic poem "*Beowulf*". In the story, brave hero Beowulf sets out with a group of warriors to slay the beast. He fights Grendel with only his bare hands and finally succeeds in tearing the monster's arm from his body. This victory is not the end of Beowulf's story, however. Read a retelling of the *Beowulf* story in this adaptation:

> *Beowulf: A Hero's Tale Retold*
> By James Rumford
> 2007

<u>Polyphemus (the cyclops)</u>

Cyclops Polyphemus is a one-eyed giant encountered by hero Odysseus in the Greek epic poem "*The Odyssey*" by Homer. Odysseus and his crew accidentally stumble into the cyclops's cave, eat their fill of his food, and fall asleep. When Polyphemus returns with his sheep, he discovers the men and traps them in his cave. Odysseus can only look on in horror as the cyclops begins to consume his shipmates. He devises a plan. First, he gets the giant drunk. When Polyphemus asks his name, Odysseus answers that his name is 'Nobody'.

The cyclops soon falls into a deep sleep and, seeing his opportunity, Odysseus blinds the giant by poking his eye with a wooden stake whose tip has been sharpened and put into the fire. When the other cyclopes on the island hear Polyphemus's cries and rush to the cave to help, the men hide. "Nobody has done this to me!" yells Polyphemus. The other giants fear that spirits have bewitched him and leave.

Odysseus and his crew attach themselves to the undersides of the giant's sheep. The next morning when the blinded cyclops opens the cave to let his sheep out, he carefully feels the back of each sheep. The men escape the cave undetected. Polyphemus never does find his archenemy, 'Nobody'.

Read more of Odysseus's adventures in these illustrated books:

Odysseus and the Cyclops: A Retelling
By Cari Meister
Illustrated by Nadine Takvorian
2012

The Adventures of Odysseus
By Hugh Lupton, Daniel Morden, and Christina Balit
2012

Ogias

The Hebrew *Bible* mentions King Og of Bashan, whose enormous iron bed was thirteen and a half feet long and six feet wide. In the *Bible*, Og is slain by Moses with a powerful strike to his ankle.

Ogias also appears in "*The Book of Giants*", an ancient Jewish text written in the 2nd century BCE. "*The Book of Giants*" tells stories of a race of giants that existed before the biblical flood. Fragments of the text were found at Qumran among the Dead Sea Scrolls.

You can read one version of Og's story in this collection of Jewish tales:

Jewish Fairy Tales and Fables
Gertrude Landa
1925

Topic 11.4: Another Binary Problem

The impression of the flower on Elizabeth's palm changes the number of rose sepals on the quilt from 85 to 90 (magic!). Convert 90 into base two (binary) and draw a picture of the new pattern of open and closed flowers.

Chapter 12

Topic 12.1: Musical intervals

Part A

When the strings of the fairy harp are plucked, vibrations are sent through the air. You can think of the vibrations as waves:

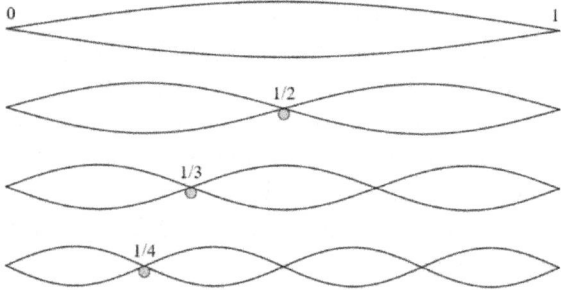

When these waves reach your eardrum, your brain interprets them as a musical note. Some sounds have a low pitch (like the sound of a tuba) while others have a high pitch (like those of a flute). The pitch of a note is determined by the frequency of the sound waves (or how many times the wave repeats in a second). In the diagram above, the long string creates a long wave. A string of half its length produces a wave that repeats two times in the same period (twice the frequency of the first wave). This shorter string will thus produce a higher pitch than the longer one.

Try this: Fill some drinking glasses with different amounts of water. Hit each one with a spoon and observe how the pitch changes with the various amounts of water.

<u>Part B</u>

A musical interval describes the ratio between the frequencies of two notes. You don't need a fairy harp to produce the musical intervals described in the book – you can use rubber bands and some math to achieve the same effect.

1. Gather the materials needed: A piece of wood, four nails, a hammer, two identical rubber bands, a ruler, and a pencil.

2. With adult supervision, hammer two nails into the wood so that when a rubber band is stretched between the nails, it is pulled taunt.

3. Hammer the other two nails so they are also the same length apart.

4. Stretch a rubber band between each pair of nails. Pluck each rubber band. They should produce the same tone.

5. Measure the distance between a pair of nails.

6. Mark the halfway point on the 2nd rubber band. Pinch the rubber band in this spot and pluck the first (full) rubber band followed by the second (pinched) rubber band. The lengths of the rubber bands are in a ratio of 2:1. This is called an octave. It should sound like the beginning of the song "*Somewhere Over the Rainbow*" from the movie "<u>*The Wizard of Oz*</u>".

7. Divide the full rubber band length by three. Mark the point on the second rubber band that is 2/3 of the full length. Pinch the second rubber band

at this point. Pluck the first (full) rubber band, followed by the second (pinched) rubber band. String lengths in the ratio 3:2 produce an interval called a perfect fifth. It should sound like the first notes of *"Here Comes the Bride"* or *"Amazing Grace"*.

8. Try producing the following musical intervals in the same manner:

Perfect Fourth: Ratio 4:3 (*"Twinkle, Twinkle"*, *"Star Wars"* theme)

Major Third: Ratio 5:4 (*"Oh When the Saints"*, *"Kumbaya"*)

Minor Third: Ratio 6:5 (*"Lullaby"* – Brahms, *"Greensleeves"*)

Major Second: Ratio 9:8 (*"Happy Birthday to You"*, *"Rudolf the Red-Nosed Reindeer"*)

Note: If you do not have access to a hammer and nails, you can stretch the rubber bands across the mouth of a bowl or box.

Chapter 13

Topic 13.1: Map coloring

The first challenge that the giants gave to Lulu was to place colored coins on a map:

- Red for Ogias
- Green for Grendel
- Blue for Polyphemus

Each coin must be placed in a single territory. The only rule regarding their placement is that no two territories can have the same colored coin if they share a border. Lulu told the giants that the task was impossible. They would need to have coins of at least four different colors.

Map coloring is a famous problem in mathematics. Let's explore it further.

What is the smallest number of colors you could use to color the following map? Stick to the rule that <u>areas of the same color may not share a border</u>. Note that a corner where two areas touch does not count as a border between them.

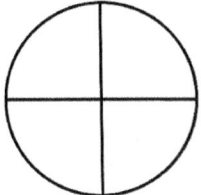

What is the smallest number of colors you would need to color the following map?

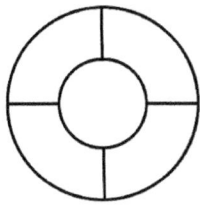

If the map Lulu was given by the giants looked like the one above, would she be able to solve the problem successfully?

Now try coloring this map using the least number of colors possible:

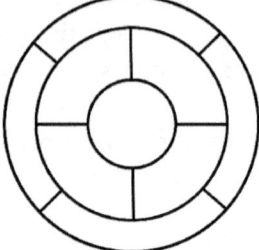

The <u>Four Color Theorem</u> states that four colors are sufficient to color any map, regardless of how complicated it is, so that areas of the same color do not share a border.

Ogias was incorrect when he said that "a theory is just a guess". In mathematics, a theorem is a statement that is not just generally accepted, but one that has also been proved to be correct. The proof of the four color theorem was accomplished using computers.

Make a photocopy of the map of England and try to color it using three colors. Is the task possible?

Make another copy of the map and color it with four colors.

261

Topic 13.2: Conic Sections

A conic section is a curve created by cutting a cone along the line of a flat plane. Form a cone out of clay and use a plastic knife (with supervision from an adult) to cut the cone into the four conic sections: circle, ellipse, parabola, and hyperbola:

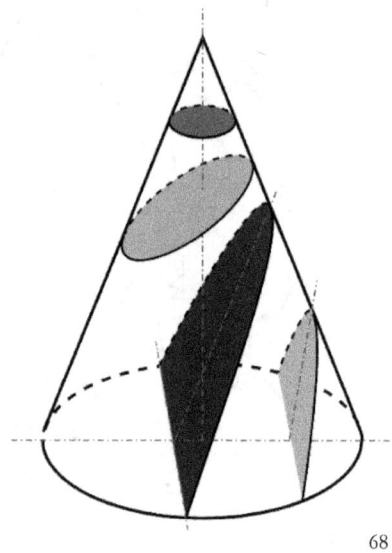

68

Topic 13.3: Monty Hall problem

Ogias showed the girls three boxes. Only one of the three contains the Holy Grail, a treasure that is a powerful relic sought by King Arthur and his knights. If they choose the correct box, the Holy Grail is theirs.

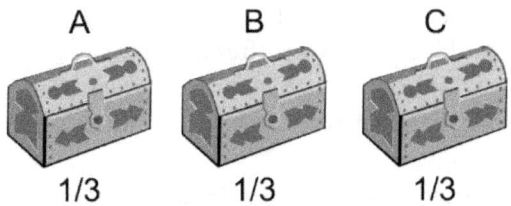

A   B   C

1/3   1/3   1/3

At the start of the game, the girls have an equal probability (a one out of three chance) of selecting the correct box. Lulu chooses Box B.

The giant opens Box A to show that it is empty. He tells Lulu that she may change her answer (select Box C) if she wishes. Should she change her answer or stick with her original guess?

Elizabeth points out that the odds have changed. How can that be? While Lulu chose at random, Ogias knows which box holds the treasure. He will always show her one of the remaining boxes that does not hold the treasure.

When Lulu selected a box, she had a one out of three (1/3) chance of being correct and a two out of three (2/3) chance of being wrong. After one of the incorrect boxes is revealed, Lulu is given the opportunity of changing her selection, it is like asking her to choose between the 1/3 option or the 2/3 option. She has better odds if she changes her selection.

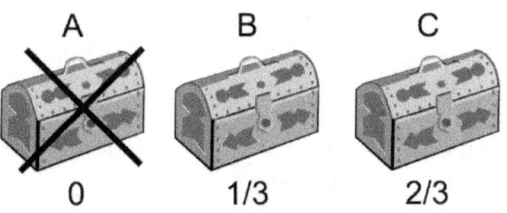

A   B   C

0   1/3   2/3

This type of probability puzzle is referred to as the "Monty Hall problem" after a game show called "*The Price is Right*" whose host Monty Hall would ask a contestant to select from three doors in the same manner as the giant's game.

If you found it hard to believe that you have better odds of winning the game after changing your answer, you are not alone. In 1990, Marilyn vos Savant explained why you should always change your answer in the Monty Hall problem to readers of the "*Ask Marilyn*" column of *Parade Magazine*. Over 10,000 readers wrote to the magazine, all adamant that her answer was incorrect.

Try out the game for yourself!

1. Gather your supplies: three opaque disposable plastic cups, a six-sided die, and one coin. You will also need a partner for this experiment. Place the cups upside down in front of you. Label the cups A, B, and C using a permanent marker.

Repeat the following steps <u>50 times</u>:

2. Turn your back while your partner rolls the die. Your partner will hide the coin underneath one of the cups as follows:

   Roll 1 or 2: Cup A
   Roll 3 or 4: Cup B
   Roll 5 or 6: Cup C

Make sure that your partner remembers which cup the coin is under.

3. Turn back to face the cups. Which cup is the coin under? Take a guess and point to one of the cups.

4. Your partner will turn over one of the remaining two cups to show you that it is empty. He/she will ask you if you would like to change your answer.

5. <u>For the first 25 trials, you will stick to your original answer. For trials 26 through 50, you will always change your answer and point to the remaining cup.</u>

6. Your partner will then reveal whether you chose correctly (if you're pointing to the cup with the coin). On a separate piece of paper, record the trial number and whether you were correct or incorrect with your final guess. Was it better to stick with your original answer or change your answer after one of the wrong answers was revealed?

Topic 13.4: Echolocation

The birds who inhabit the giants' cave are swiftlets. Aerodramus swiftlets (found in tropical climates) are unique in their ability to use echolocation to "see" in dark caves. They send out a series of clicks and listen to the sound of their echoes as the sound waves bounce off objects in their environment. Bats, dolphins, and whales also use echolocation.

Humans (both blind and seeing) can perform basic echolocation with some training. Try this experiment:

<u>Part A</u>

1. Locate a blindfold and a partner for the experiment.

2. Find a large room to use for the experiment. It is preferable for the room to be carpeted so your partner can move around the room without detection. Place a chair in the center of the room

3. On a separate piece of paper, draw a basic diagram of the room. Make sure you include the chair in the center.

4. Put on the blindfold and sit down on the chair.

5. Ask your partner to go somewhere in the room and make a short clicking sound.

6. Without leaving your chair, point to the location from which you heard the sound.

7. Your partner should mark the location on the paper and indicate whether or not you were correct.

8. Repeat steps 4-7 ten times using different spots in the room (in front of you, behind you, to either side, near, far, etc..).

9. Analyze the results. How did you do?

<u>Part B</u>

Learning to echolocate takes hours of training. Try this to get started:

1. Stand in the center of a room and close your eyes.

2. Click your tongue and listen for the sound it makes and the echo that is created.

3. Now stand one foot away from a wall.

4. Close your eyes and click your tongue again. Listen to the sound. How is it different than the sound/echo when you were standing in the center of the room?

5. Repeat, this time standing one foot away from the corner of the room.

6. Keep practicing in these three locations until you feel you can tell them apart by the sounds echoing from your clicks.

7. Put on a blindfold and ask a friend to guide you to one of the three locations (he/she may need to first spin you around, so you don't know where you are being led).

8. Click your tongue and try to identify your location. How did you do?

Chapter 14 (answers on page 280)

Topic 14.1: Jabberwocky

Jabberwocky is a creature imagined by Lewis Carroll in his nonsense poem of the same name. The poem Jabberwocky is included in Carroll's novel _Through the Looking-Glass, and What Alice Found There_ (the sequel to _Alice in Wonderland_). Look for this picture-book rendition of Carroll's poem, illustrated by the talented Graeme Base:

Jabberwocky
By Lewis Carroll
Illustrated by Graeme Base
2000

Topic 14.2: When and Where Will the Battle Begin?

The giants asked Elizabeth to calculate the time and the distance from Camelot where the Black Pig and King Hallden will meet and the war will commence. She refused to reveal the answer and help the dark side with their battle plans. Can you solve the problem? Here are the clues given by Ogias:

1. The Black Horse Vale is 115 km away from Camelot castle.

2. At precisely eight o'clock, the Black Pig will ride straight toward Camelot at 60 kilometers per hour.

3. He expects to be spotted as he passes Hengoen Hill, which is forty-five kilometers into his trip.

4. Fifteen minutes later, King Hallden will come riding to meet him - at a rate of 50 km per hour.

5. When they meet, the Black Pig will give the signal that the war has begun. When and where will this occur? (keep reading for the answer)

These types of math problems are often called "train speed" problems because instead of knights riding at each other, they frequently involve trains. The most important thing to know in problems like these is:

$$Distance = Rate \times Time$$

A rate of 60 km per hour just means that the horse will cover 60 km in one hour, 120 km in two hours, etc.

Sometimes it is easier to work with minutes rather than whole hours. If the Black Pig is riding at 60 km/hr. and there are 60 minutes in an hour, he is traveling at a rate of 1 km every minute. The Black Pig will reach Hengoen Hill at 8:45 (as it is 45 km from his starting point).

It will take King Hallden 15 minutes to receive the warning call and suit up for battle. He will leave Camelot at exactly 9:00. At this point, the Black Pig had 15 more minutes of riding, so he will be 60 km away from his starting position and 55 km away from Camelot Castle (115-60).

King Hallden rides at a rate of 50 km per hour. In half an hour, he will travel 25 km. During the same period, the Black Pig will travel 30 km. 30+25=55, which is the rest of the distance. They will clash at 9:30 pm at a location 25 km from Camelot.

Try this additional problem:

Fifteen minutes after the Black Pig departs, the new Fairy Queen, Nimue, arrives at Black Horse Vale. She enters Merlin's tent, and they speak in hushed tones for another quarter of an hour.

Without explanation, Merlin suddenly races from his tent and immediately mounts a supernatural beast. Could it be that Nimue successfully convinced him of the errors of his ways? Does he intend to stop the battle or is he bringing more dark magic to aid the Black Pig?

Merlin sets off in the same direction as the Black Pig. For Merlin to overtake the Black Pig before he reaches King Hallden, at what rate must his beast travel?

Chapter 16

Topic 16.1: Rudyard Kipling

English poet and novelist Rudyard Kipling (1865-1936) was born in Bombay to British parents. At five years old, he was sent away from his family to attend school in the United Kingdom. Shortly before his 17th birthday, Kipling returned to India, a land which was a source of inspiration in his writing.

Many of Rudyard Kipling's poems are integrated into his novels and short stories such as _The Jungle Book_ and _Just So Stories for Little Children_. You can also look for this illustrated collection of Kipling's poems in your local library:

> _Rudyard Kipling_
> _(Poetry for Young People Series)_
> By Rudyard Kipling,
> Edited by Eileen Gillooly,
> Illustrated by Jim Sharpe
> 2000

Topic 16.2: Poem Code

In his book, _Between Silk and Cyanide: A Codemaker's War 1941-45_, Leo Marks describes some of the cryptography techniques used in World War II by the Special Operations Executive (SOE) which he headed.

One of these techniques is the 'poem code'. Here's how it works:

1. In a poem code, certain words in the poem act as the key to decrypting the message. In the message from Mrs. Magpie, the word "WONDERLAND" is the key. Typically, an additional code at the beginning of the poem indicates which words comprise the key.

2. The first step is to assign numbers to the letters in the key. Since WONDERLAND has ten letters, we will assign the numbers from 1 to 10 to those letters based on the order they appear in the alphabet. A is assigned 1. Since the letters B and C do not appear in the key, the D on the left is assigned 2 and the next D is assigned 3. Continue until all the letters in the key are assigned numbers:

W	O	N	D	E	R	L	A	N	D
10	8	6	2	4	9	5	1	7	3

3. The number of letters in the key is the number of columns in the final message. The numbers assigned to the letters show you the order in which you must write the encrypted message in the columns. For example, the first group of letters goes in the 10th column, the next group of letters goes into the 8th column, the following group in the 6th column, and so on.

Note: Mrs. Magpie has simplified this process. The encrypted message usually doesn't have

commas between the letters for each column.
You would divide the number of total letters
by the number of columns to figure out how
many letters each column holds.

uhmendig, wretdree, nfdoadTi, usonpoaii, aulunsts,
btwhueCa, stattihr, BdTescEta, eonfnehs, tpgOenren

1	2	3	4	5	6	7	8	9	10
B	u	t	a	s	n	e	w	b	u
d	s	p	u	t	f	o	r	t	h
T	o	g	l	a	d	n	e	w	m
e	n	O	u	t	o	f	t	h	e
s	p	e	n	t	a	n	d	u	n
c	o	n	s	i	d	e	r	e	d
E	a	r	t	h	T	h	e	C	i
t	i	e	s	r	i	s	e	a	g
a	i	n							

It is up to you to put the appropriate spaces in the
message:

> *But, as new buds put forth*
> *To glad new men,*
> *Out of the spent and unconsidered Earth,*
> *The Cities rise again.*

Answers

Chapter 1

1.1. The Pigeon's journey will take him 36 hours (2,700 divided by 75). It would take him a full day (24 hours) plus 12 additional hours. The message was delivered at 10 am on Sunday, so 36 hours later would put the Pigeon's arrival at:
10 pm on Monday.

Chapter 2

2.1. Rhyming scheme matching:

1. AABCCB –
   "At the Sea-Side" (Robert Louis Stevenson)

2. ABAAB -
   "*The Road Not Taken*" (Robert Frost)

3. AABBA –
   A Book of Nonsense (Edward Lear)

2.6.A. If the book was published in 1976 and the maximum copyright is 95 years, it will enter the public domain in the year 2071 (calculate 1976 + 95).

Chapter 4

4.1. The twins started at 49 N, 77.47 W and traveled first to the outpost at 50 N, 40 W. They traveled 1 degree North and -37.47 West (or 37.47 East). If the second half of their journey covers this same distance (in the same direction), they will end up at 51 N, 2.53 W. At coordinates 51.02456N, 2.531634W, you will find Cadbury Castle in England. Many people believe that this castle (formerly named Camalet) is the site of King Arthur's Camelot.

4.2.A.
> 1. 10 feet
> 2. 30 knots
> 3. 30 nautical miles

Chapter 5

5.3. In addition to three sepals and three petals, the lily has six stamens (a multiple of three), as well as a three-parted stigma. If you cut horizontally across the ovary of the lily, you will see that the ovules are attached at three points.

5.4. Part B.
> 1. A: "has a"
> 2. B: "is a"
> 3. B: "is a"
> 4. A: "has a"
> 5. A: "has a"
> 6. B: "is a"

5.5. Part A

Button 3	Button 2	Button 1	Binary Base-2	Value Base-10
●	●	●	000	0
●	●	✿	001	1
●	✿	●	010	2
●	✿	✿	011	3
✿	●	●	100	4
✿	●	✿	101	5
✿	✿	●	110	6
✿	✿	✿	111	7

You can represent eight values (from 0 to 7) using three buttons. With four buttons, you can represent 16 values, and with five buttons, you can represent 32 values.

5.5 Part B.

A. 8 in base-2 is 1000
$(8=1\times8+0\times4+0\times2+0\times1)$

B.30 in base-2 is 11110
$(30=1\times16+1\times8+1\times4+1\times2+0\times1)$

C. 52 in base 2 is 110100
$(52=1\times32+1\times16+0\times8+1\times4+0\times2+0\times1)$

Chapter 6

6.3. These types of logic problems are often called "Missing Dollar Riddles" because they often involve money rather than gemstones. The problem attempts to trick the reader by adding up unrelated numbers.

Week	Gems Removed	Queen Gems	Nimue Gems
0	-	20	0
1	6	20-6=14	0+6=6
2	6	14-6=8	6+6=12
3	8	8-8=0	12+8=20

Adding the number of gems in the Fairy Queen's possession is irrelevant – the total number of gems can be calculated by looking at how many were removed each week: 6+6+8=20. There were never any additional gems created – poor Nimue was tricked!

Chapter 7

7.1.

   A. Three socks will guarantee a pair,

   B. Five socks will guarantee two pairs (with four socks, you could get three of one color and one of another)

   C. There are 365 days in a typical year, but 366 days in a leap year (so 366 different birthdays a person could have). Inviting 367 people to the party will guarantee that at least two will share a birthday.

   D. Try the "hand-shaking problem" with a smaller number. If there were 2 people at the party, you can't have one person shake hands with no people and the other shake hands with only one. If you have 3 people, one person can shake hands with zero people, but the other two must shake hands with each other (thus shaking hands with the same number of people). Even if you keep adding people to the party, it is impossible for each person to shake hands with a different number of individuals.

7.5. A.

		Heterozygous Round (Rr)	
		R	**r**
Heterozygous Round (Rr)	**R**	RR	Rr
	r	Rr	rr

Mendel discovered that one-fourth of the offspring had wrinkled peas.

7.5. B. With seven traits and two values for each trait, you would have 2 x 2 x 2 x 2 x 2 x 2 x 2 = 2^7 = 128 combinations of traits.

7.5. C. All the pea plants in the second generation will be YyRr. Their peas will be yellow and round.

7.5. D.

		YyRr			
		YR	**Yr**	**yR**	**yr**
YyRr	**YR**	YYRR	YYRr	YyRR	YyRr
	Yr	YYRr	YYrr	YyRr	Yyrr
	yR	YyRR	YyRr	yyRR	yyRr
	yr	YyRr	Yyrr	yyRr	yyrr

Yellow and Round: 9 / 16
Yellow and Wrinkled: 3 / 16
Green and Round: 3 / 16
Green and Wrinkled: 1 / 16

Chapter 9

9.2. Fill the 4-cup measuring cup and pour its contents into the 7-cup measuring cup. The 7-cup measuring cup can hold 3 more cups of flour (7-4=3). Fill the 4-cup container again and pour into the 7-cup container until that cup is full. Now you have 1 cup left in the 4-cup measuring cup (4-3=1). Pour it into your mixing bowl.

Fill the 4-cup measuring cup one more time and pour into the mixing bowl. You've added 1+4=5 cups of flour to your recipe.

9.4.

B	E	W	A	R	E
— ●●●	●	● — —	● —	● — ●	●

Chapter 10

<u>10.1</u>

There were at least 23 eggs in the basket:
(3×7)+2=23
(5×4)+3=23
(7×3)+2=23

Chapter 11

<u>11.1</u> Move these three toothpicks (and the eye):

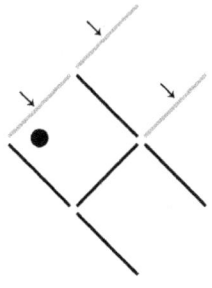

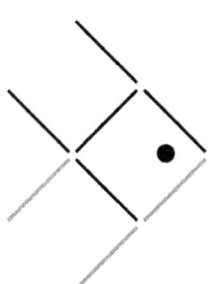

<u>11.2</u>

The factors of 12 are 1, 2, 3, 4, 6, and 12. Their sum is 28, so the abundancy is 28÷12=7/3.

11.4

The number 90 in base-2 (binary) is: 1011010

64+0+16+8+0+2+0=90

64	32	16	8	4	2	1
✹	●	✹	✹	●	✹	●

Chapter 14

14.2. Merlin's ride begins 30 minutes after the Black Pig's. In the main problem for this chapter, we calculated that the Black Pig would clash with Kind Hallden an hour and a half into his 90-km journey (115-25=90). Merlin would have one hour to ride 90 kilometers.

Remember: Distance = Rate × Time

Merlin's rate of speed would have to be 90 km/hr. A horse would not be able to accomplish this, but perhaps it is possible for a supernatural beast.